L'HOMME ET LA NATURE

Actes de la Société canadienne d'étude du dix-huitième siècle

Tome III

publié sous la direction de
Robert James Merrett

* * * * * *

pour la Société
par
Academic Printing and Publishing
Edmonton, Alberta
1984

MAN AND NATURE

Proceedings of the Canadian Society for Eighteenth-Century Studies

Volume III

Edited by

Robert James Merrett

* * * * * *

Published for the Society
by
Academic Printing and Publishing
Edmonton, Alberta
1984

Man and Nature — ISSN 0824-3298

Volume III — ISBN 0-920980-04-X (*cloth*)
0-920980-05-8 (*paper*)

Printed and published in Canada by
Academic Printing and Publishing
P.O. Box 4834
South Edmonton, Alberta, T6E 5G7

Contents/Table des matières

Acknowledgements/Remerciements

Many people have helped to bring about this third volume of *The Proceedings of the Canadian Society for Eighteenth-Century Studies*. The Society and the editor owe a debt of gratitude to the following assessors of articles: Professors Roger Clark (Memorial), David Jory and Charles Stewart-Robertson (both of UNB at Saint John), and Victor Chan, Joan Crowther, Greg Hollingshead, David Jackel, and Ted Pitcher (all of Alberta). Professor Ed Greene (Alberta), besides serving as an assessor, gave the editor invaluable encouragement during the preparation of the volume, as did Professor Roger Emerson (Western Ontario). It is a pleasure to acknowledge grants in aid of publication from Dr. J.G. Kaplan, Vice President (Research) at the University of Alberta and from the Social Sciences and Humanities Research Council of Canada. Lastly, the editor would like to thank Dr. Roger Shiner of Academic Printing and Publishing for his cordial assistance with the publication of this volume.

* * *

Beaucoup de personnes ont contribué à réaliser ce troisième volume des *Travaux de la Société Canadienne d'Etudes du Dix-huitième Siècle*. La Société et l'éditeur reconnaissent la dette de gratitude qu'ils doivent aux assesseurs d'articles suivants: les professeurs Roger Clark (Memorial), David Jory et Charles Stewart-Robertson (les deux de l'UNB à Saint John), Victor Chan, Joan Crowther, Greg Hollingshead, David Jackel et Ted Pitcher (ses cinq de l'U. de l'Alberta). Le Professeur Emérite Ed Greene (Alberta) a servi comme assesseur tout en donnant à l'éditeur du volume un encouragement inestimable tout comme le Professeur Roger Emerson (Western Ontario). C'est un plaisir de reconnaître des donations de la part du Dr. J.G. Kaplan, Vice-Président (Recherches) de l'Université de l'Alberta, et aussi du CCRSSH. Enfin, l'éditeur désire remercier le Dr. Roger Shiner de la maison d'édition *Academic Printing and Publishing* pour l'avoir aidé si cordialement dans la publication de ce volume.

Preface/Préface

The nine essays which have been selected and prepared especially for this volume started out as papers that, along with twenty-nine others, were read at the ninth general meeting of the Canadian Society for Eighteenth-Century Studies held at the University of Alberta on October 14-16, 1982. No common theme ties the essays together. They deal with European culture, critical history, major English and French authors, the status of theatrical texts, Quebec colonial history, the sense of ending in German drama, the problems of canonic art, and explicit and implicit symbolic meaning in music. However, besides representing a wide range of the Society's disciplines, the essays as a group contribute to our understanding of the Eighteenth Century because they have in common a historical sense of the tension between tradition and innovation. In their different ways, the nine essayists challenge single-minded views of the period, preferring to emphasize, not doctrinaire polarities, but vital conflicts.

In his study of vampires, Milan Dimić explores the Age of Reason's fascination with superstition. He demonstrates that throughout Europe, in courts, academies and journals, there was a preoccupation with the daemonic which was both exploited for satirical purposes and converted into high literature, namely, Goethe's 'The Bride of Corinth.' Dimić's comparatist study defends the view that there is a tension between reason and emotion in the whole course of the Eighteenth Century. By explaining the gap between the cultivated use of allusion in English literature and the lack of a theory of allusion, William Kinsley is also concerned to study tensions within literary history. Discussing contemporary views of imitation, Kinsley points out the increasing anxiety writers felt about what is now referred to as intertextuality. His major contention is that the period's lack of critical concepts about allusion, since it stems from an intolerance regarding the paradoxical relation of texts, cannot be allowed to limit description of eighteenth-century literature, especially of the allusive intuitions of Dryden and Pope.

Mark Madoff's presentation of the sartorial trope in *Gulliver's Travels* modifies the traditional notion that satire is a hodge-podge. By proving the trope to be ubiquitous, Madoff shows that it provides Swift's work with a topical unity. The tailoring and the wearing of clothes is a recurrent analogy for the making and the possessing of philosophical ideas. Whereas Madoff freshly discusses the imagistic and topical stability of *Gulliver's Travels,* Roseann Runte's new perspective on Voltaire's 'Zadig' emphasizes how much the story resists interpretation by perpetually subjecting opposing determinist and optimistic hypotheses to contradiction. Voltaire concentrates on addressing the reader's critical intelligence by obliging the reader to recognize that the text of the story is deeply antithetical. Like Runte, David Trott relies on deconstructionist assumptions. He stresses the experimental nature of eighteenth-century French theatre. Because Fuzelier wrote multiple versions of his dramatic texts, he affords Trott the opportunity to describe the serial processing of these versions and to propose a model of intertextuality which reveals the need for a new theory of dramatic texts.

Catherine Rubinger's essay on Mme Bégon's correspondence accentuates the similarities between the salons of France and Quebec in order to highlight the comprehensive domestic, social, and cultural roles available to women in New France. Although Mme Bégon's impromptu salons did not deal in high-minded fashion with literature or philosophy, they permitted her to see herself at the centre of colonial life and to regard hospitality as a calling. While a more literary essay, Anna Wittmann's piece on dénouement in German plays also makes an important contribution to cultural history. Wittmann classifies eighteenth-century dénouements to de-emphasize the radical nature of Romantic drama and to stress the continuity of dramatic form. She demonstrates that eighteenth-century closures, in that they reveal a reduced dependence upon socially prominent characters, poetic justice and benevolence, lay the foundation of Romantic dénouements. In his examination of Mengs' theory and practice, Calvin Seerveld similarly deals with the tension between tradition and innovation; he maintains that this tension accounts for Mengs' understanding of canonicity. Committed to the classical idea of imitation, Mengs also highly valued the practical rules of art. Holding to the model of Greek art and believing modern artists can perfect the masters, Mengs, according to Seerveld, profitably fathomed the problems associated with art canons. Although Jacobus Kloppers provides a compelling historical context for Bach's organ music, he does not thereby diminish its universal appeal. If, on the one hand, by illustrating Bach's traditional use of rhetoric and symbols, Kloppers evidences the composer's resistance to naturalistic concepts of music, on the other hand, he proves that Bach's music integrates rational and emotional values. Relying on transparent as well as coded symbols, Bach's organ music is not only a vehicle for religious ideas; it is in itself spiritual.

If one virtue of such a volume is that it affords the reader interdisciplinary insights, as, for example, the opportunity to compare the essays of Seerveld and Kloppers from the point of view of their different but complementary treatments of the sister arts, it is also valuable because it invites the reader to ponder the continuities of literary change, to reflect on the importance of refining critical theory, and to appreciate the vital tensions between tradition and innovation in eighteenth-century culture.

* * *

Les neuf essais choisis et préparés spécialement pour ce volume sont neuf parmi trente-huit présentés à la neuvième réunion générale de la Société Canadienne d'Etude du Dix-huitième Siècle à l'Université de l'Alberta les 14-16 octobre 1982. Nul thème commun ne relie ces essais qui traitent de la culture européenne, de l'histoire de la critique, d'auteurs majeurs anglais et français, de l'état de textes dramatiques, de l'histoire coloniale du Québec, du sens de la fin dans le théâtre allemand, des problèmes de l'art canonique et du sens symbolique explicite et implicite dans la musique. Ces neuf essais représentent donc la large gamme des disciplines de la Société et contribuent à notre compréhension du Dix-huitième Siècle parce qu'ils ont en commun le sens historique de la tension entre la tradition et l'innovation. Les neuf essayistes, chacun de sa façon personnelle, défient les vues bornées de la période et préfèrent mettre en relief, non les polarités doctrinaires, mais les conflits vitaux.

Dans son étude des vampires, Milan Dimić explore la fascination qu'avait pour la superstition l'Age de la Raison. Il démontre que partout en Europe, dans les cours, les académies, les journaux, on était préoccupé du démoniaque soit pour exploiter des buts satiriques soit pour atteindre la haute littérature, notamment dans *La Fiancée de Corinthe.* L'étude comparatiste de Dimić souligne l'idée qu'il y a une tension entre raison et émotion dans tout le cours du Dix-huitième. En expliquant la brèche entre l'emploi cultivé de l'allusion dans la littérature anglaise et le manque d'une théorie de l'allusion, William Kinsley tient à étudier les tensions dans l'histoire de la littérature. En discutant les idées contemporaines de l'imitation, Kinsley souligne l'anxiété croissante des auteurs au sujet de ce qu'on appelle aujourd'hui 'l'intertextualité'. Son affirmation principale, c'est que le manque de concepts critiques sur l'allusion pendant cette période, puisqu'il dérive d'une intolérance concernant la relation paradoxale des textes, ne doit pas limiter la description de la littérature du Dix-huitième, surtout les intuitions allusives de Dryden et de Pope.

La présentation de Mark Madoff du trope sartorial dans les *Voyages de Gulliver* modifie la notion traditionnelle que la satire est un méli-mélo. En prouvant que le trope est omniprésent, Madoff montre qu'il donne à l'oeuvre de Swift une unité topique. Le fait de fabriquer et de porter des vêtements est une analogie qui revient pour la fabrication et la possession d'idées philosophiques. Tandis que Madoff apporte du nouveau sur la stabilité imagiste et topique des *Voyages de Gulliver*, la nouvelle perspective dans laquelle Roseann Runte voit le *Zadig* de Voltaire met en relief combien le conte défie l'interprétation en soumettant toujours à la contradiction des hypothèses déterministes et optimistes. Voltaire s'adresse surtout à l'intelligence critique du lecteur en l'obligeant à reconnaître que le texte du conte est profondément antithétique. Comme Runte, David Trott se fie à des suppositions déconstructionistes. Il insiste sur la nature expérimentale du théâtre français du Dix-huitième. Puisque Fuzelier écrivait de multiples versions de ses textes dramatiques, il donne à Trott l'occasion de décrire le développement en série de ces versions et de proposer un modèle d'intertextualité qui montre le besoin d'une nouvelle théorie de textes dramatiques.

L'essai de Catherine Rubinger sur la correspondance de Mme Bégon souligne les ressemblances entre les salons de France et du Québec afin de mettre en relief les rôles compréhensifs à la fois domestiques, sociaux et culturels ouverts aux femmes de la Nouvelle France. Bien que les salons impromptus de Mme Bégon ne traitaient pas de façon présomptueuse de littérature ou de philosophie, ils lui permettaient de se voir au centre de la vie coloniale et de considérer l'hospitalité comme une vocation. Une communication plus littéraire, l'essai d'Anna Wittman sur le dénouement dans le théâtre allemand, apporte aussi une contribution importante à l'histoire de la culture. Sa façon de classer les dénouements du Dix-huitième réduit la nature radicale du drame romantique et insiste sur la continuité de la forme dramatique. Elle démontre que les dénouements du Dix-huitième, en révélant une moindre dépendance de personnages d'importance, de la justice poétique et de la bienfaisance, fondent les dénouements du Romantisme. Dans son examen de la théorie et de la pratique de Mengs, Calvin Seerveld lui aussi traite de la tension entre tradition et innovation; il soutient que cette tension explique la façon de Mengs de comprendre la canonicité. Fidèle a l'idée classique de l'imitation, Mengs prisait surtout les règles pratiques de l'art. Soutenant le modèle de l'art grec et croyant que les artistes modernes peuvent devancer les maîtres, Mengs selon Seerveld a su pénétrer utilement les problèmes associés aux canons de l'art. Bien que Jacobus Kloppers nous donne un contexte historique irrésistible pour la musique d'orgue de Bach, il ne diminue nullement son attrait universel. Si, d'une part, en illustrant l'emploi traditionnel par Bach de la rhétorique et des symboles, Kloppers met en évidence la résistance du compositeur aux concepts naturalistes de la musique, de l'autre côte il démontre que la musique de Bach coordonne les valeurs rationalistes et émotives. Faisant fond sur des symboles transparents aussi bien que les codés, la musique d'orgue de Bach n'est past seulement un véhicule pour des idées religieuses, elle est elle-même spirituelle.

Si une qualité d'un tel volume c'est qu'elle offre au lectueur des aperçus interdisciplinaires comme, par exemple, l'occasion de comparer les essais de Seerveld et de Kloppers du point du vue de leurs façons différentes mais complémentaires de traiter des arts soeurs, le volume invite aussi le lecteur à réfléchir sur les continuités des changements littéraires et à considérer l'importance de raffiner la théorie critique et d'apprécier les tensions vitales entre la tradition et l'innovation dans la culture du Dix-huitième Siècle.

ROBERT JAMES MERRETT
University of Alberta

1. Vampiromania in the Eighteenth Century: The Other Side of Enlightenment

Strange shadows persisted in the period which is commonly called the Enlightenment, *le Siècle des Lumières, Aufklärung, illuminismo, siglo de las Luces, seculo das Luzes, prosveščenie*, or *prosvečenost*. The Age of Reason was not only the time of the philosophers, neoclassical rules and certified traditions. It was also the era of rococo decoration and architecture, Piranesi's 'Prisons'; of gloomy nights, haunted churchyards, and desolate ruins in the poetry of Young, Gray, Thomson, and Gessner; of the tortured sensibility of Richardson's novels, the anguish of *Manon Lescaut*, the *Nouvelle Héloise* and *Werther;* of the integral originality of the *Sturm und Drang* group and of Sebastian Mercier; of the 'crisis of laughter' which began with the black humour of Swift's *Modest Proposal;* of Macpherson's 'Ossian' and the rediscovery of genuine ballads, tales and legends of folklore; of nightmarish Gothic romances, dark tales of brigandism, and de Sade's reversal of codes and morals; of Germanic pietism, Wesley's methodism, and Fénelon's and Rousseau's personalized catholicism of feelings; and it was also the time of the esotericism and illuminism of secret societies ... .[1]

Intense feeling for the holy and the daemonic persisted throughout the period. Controversies over miracles, divination, witchcraft, and possession by evil spirits swept Europe and some of the colonies.[2] Probably the most fashionable and the most scandalous irrational element was the fear of living corpses: those who have perished more than once, have known

0824-3298/84/0300-0001 $01.25/©C.S.E.C.S./S.C.E.D.S.

the land of darkness and the shadow of death, and who have still refused to conform to Pascal's maxim 'On mourra seul.' The troublesome walking dead belonged, of course, to the beliefs and superstitions of all times and nations. Preternatural beings that are said to suck the blood of persons at night seem to have been known in most regions of the world at least since the times of Sumer, Babylon, Assyria, the Egypt of the pharaohs, and the Jews of the Old Testament, as well as in the younger civilizations of India, China, and Japan, and in the Graeco-Roman and Arabic cultures. In literature, the most famous infamous feeders on blood, among many ancient ones, have been the Indian and Arabic *ghoul*, the Greek λαμία, and Roman *larva.*[3]

In the Middle Ages, rulers and church dignitaries often had to forbid the opening of graves and killing of corpses, regardless of the deceased's reputation. Such prohibitions were made or mentioned by Charlemagne in the eighth century, by Bishop Burchard of Worms around 1000, the Second Council of Limoges in 1031, the theologian Guillaume d'Auvergne in the first half and the Dominican judge Etienne de Bourbon in the middle of the thirteenth century, and by the Serbian emperor Stefan Uroš Dušan Nemanjić in the middle of the fourteenth. Fairly clear cases of the un-dead who attacked the living were described by William of Newburgh, who under the year 1196 notes in his chronicle the case of a 'sanguisuga.' Such episodes appeared also in various Icelandic sagas, which were usually recorded about the middle of the thirteenth century. The oldest Slavic stories appeared in a Czech chronicle of the fourtheenth-century, but are preserved only in a dubious manuscript of 1587. In the Balkans, there are reports in Latin, German, and the native tongues dating from the fifteenth, sixteenth and especially the seventeenth centuries. During the sixteenth century, such anecdotes were related in Czechoslovakia, Poland, and particularly in German speaking countries, so that Martin Luther once had to reassure a fainthearted priest from Wittenberg that they were 'des Teufels Betrügerei und Bosheit; wenn sie es nicht gläubeten, so schadete es ihnen nicht' (Tischrede Nr. 6823, Weimarer Ausgabe).

A few of the seventeenth-century rumors from many corners of Europe were printed in a number of languages. For some 'incidents,' up to twenty-three versions and re-editions have been identified, for example those in François de Rosset's *Les Histoires tragiques de notre temps* (1605) and its German adaptation by Martin Zeller (1625). Corpses able to kill and to die a second time appeared also in such popular German anthologies of magical phenomena as Henning Gross' *Magica* (1600) and especially Erasmus Franciscus' *Der höllische Proteus* (1690 and 1708). A German scholar from Leipzig, Philipp Rohr, published the first scientific investigation in 1679, the *Dissertatio historico-philosophica de masticatione mortuorum;* he rejected some cases as unverified and explained others through natural causes, but concluded that a few must be the workings of the devil and of warlocks. Among serious writers, such beliefs are mentioned by the Cambridge Platonist Henry More (*An Antidote Against Atheism*, 1653, b. III, ch. viii-ix), Paul Rycaut (*Histoire de l'état présent de l'église grecque et de l'église arménienne*, 1692), and Joseph Pitton de Tournefort (*Voyage au Levant*, 1702). Two of the most startling revelations appeared in the French journal *Mercure Galant* in May 1693 and February 1694, which is exactly at the time deemed by Paul Hazard to be the critical period of the enlightenment's 'crise de la conscience européenne.'[4] The first contribution, under the title 'Article fort extraordinaire,' and the second, under the heading 'Traité sur les Striges de Russie,' described vampiric phenomena in Russia and parts of Poland, attributing the crimes to the devil or the body soul. The filthy habits and the heretical creed of these barbaric peasants have provoked such unusual visitation by a just Lord. These anonymous articles were often reprinted in the eighteenth century, the reprints unobtrusively substituting the terms *vampire* and *oupire* for the circumlocutions used in the original. In any case, after Jan Sobieski's victory over the Turks at Vienna in 1683 and the subsequent advances of the Austrian armies into the Balkans, the time seemed ripe to show greater interest in the backward and sinful ways of the Slavic tribes.

Delegated by Emperor Charles VI (Karl VI) as official Austrian witnesses, Frombald, Glaser, and later Flückinger described in 1725 the antics of Petar Plogojević (probably Blagojević) in the village Kisiljeva and even more dramatically in 1731/32 the crimes of the Arnautin (i.e. Albanian) Pavle (the Arnold Paole or Paul of Western sources) in the hamlet Medveđa, very small places not far from Belgrade. These reports launched the term *vampire* (or *vampyre*), a word of Slavic origin whose beginnings are not known (from *upyr*, blood-sucker?) or, less likely, derived from Turkish words for wizard and witch (*über, ubęr).*[5] In 1725, the news made head-lines in fashionable journals like *Wiener Diarium*, *Vossische Zeitung*, *Leipziger Zeitungen*, *Breßlauische Sammlungen*, and was publicized in broadsheets (the so called 'Fliegende Blätter'). In 1732, the year of the great vampire scare, articles appeared twice in *Le Glaneur Historique*, repeatedly in *Mercure de France*, and once in *Mercure Historique et Politique*, *The London Journal*, *The Gentleman's Magazine*, and *The Craftsman*. In German speaking countries twelve books and four dissertations appeared in this year and the next. Some books were the result of the labours of more than one author, and one specialist, Johann Christian Harenberg, wrote three volumes in seven years, but all the works concerned stressed in their titles the sensational and bloody nature of the events, as well as the Christian and reasonable character of their own interpretation. Quotations and summaries found their way into standard encyclopaedias, like *Zedlers Universal Lexikon* (v. XLVI, 1745), and into popular works on other topics, for example the *Lettres juives* by the Marquis Boyer d'Argens (lettre 137, beginning with the 1738 edition, and also in the English and German translations), *The Harleian Miscellany: A Collection of Scarce, Curious and Entertaining Pamphlets and Tracts, As Well in Manuscript as in Print, Found in the Late Earl of Oxford's Library* (v. IV, 1745, sect. i-ii, pp. 348-59, often reissued), and, most particularly, into Dom Augustin Calmet's notorious *Dissertations sur les apparitions des anges, des démons et des esprits et sur les revenants et vampires de Hongrie, de Bohémie, de Moravie et de Silésie* (Paris, 1746).[6] This latter work by the well-known French Biblical scholar and church historian was reprinted in 1749, and a revised edition published in 1751 under the title *Traité sur les apparitions* ... ; it was translated into major European languages and read by most leading intellectuals of the century. The book contained an almost complete survey of all 'well attested cases.'

At this point it is useful to freely translate into English a few passages of the original report:

Having heard from various quarters that ... so-called vampires have been responsible for the death of several persons, by sucking their blood, I have been commissioned by Her Majesty to throw some light on this question. ... This report has been compiled with the help of the captain of the company of *heyduks* [foot soldier], and other most respected *heyduks* of the neighbourhood. After much questioning, these have declared unanimously that *about five years ago a heyduk of the area, named Arnold Paole, broke his neck, falling from a hayrick:* the said Arnold Paole had told various people, in the course of previous years, that he had been bitten by a vampire, near Cossowa, in Turkish Serbia.

This is why he himself ate earth taken from the tomb of a vampire, and bathed his wounds in the blood of a vampire (as is the custom) to "cleanse" himself of its cursed influence. However, twenty or thirty days after his death, several persons complained that the said Arnold Paole had come back to torment them, and that he had caused the deaths of four others. In order to put a stop to this danger, the *heyduk* suggested that the vampire be disinterred: this was duly done, forty days after his death, and he was found to be perfectly preserved. His flesh had not decomposed, his eyes were filled with fresh blood, which also flowed from his ears and nose, soiling his shirt and funeral shroud. His fingernails and toenails had dropped off, and others had grown through in their place, from which it was concluded that he was an arch-vampire. So, according to the custom of those regions, a stake was driven through his heart. ...

He gave a great shriek, and an enormous quantity of blood spurted from his body. The body was burned that same day and the ashes cast into his tomb. But the people in these parts claim that all those who are victims of a vampire become vampires in turn when they die. That is why it was decided to execute the four corpses mentioned above, in the same way. ...

The *heyduk* Jobira let it be known that his daughter-in-law Stanojka, having gone to bed a fortnight before, in perfect health, had woken up screaming horribly; terrified, she claimed that she had been *touched on the neck* by a man who had been dead for more than four weeks, the son of *heyduk* Miloje. From that moment, she became weaker and weaker and died within eight days. ...

Of the twelve exhumed corpses, eight were 'judged to be in a vampiric state'; the rest had decomposed. Stanojka still bore the marks of the night visitor: *'under her left ear, we could clearly make out a bluish scar, from which blood had been sucked.'* The report, like many others, coyly alludes to the sexually aroused state of some of the male vampires.

After having taken the appropriate action, we ordered the heads of all these vampires to be cut off by some wandering Bohemians, their bodies to be burned, and their ashes scattered in Moravia, while the corpses found to be in a state of decomposition were returned to their coffins. I affirm — together with the assistant field surgeons despatched to me — that all these things took place just as we have reported them at Medwegya, in Serbia, on January 7th 1732. Signed: Johannes Flückinger, Regimental Surgeon [and four others].[7]

The sensational news was earnestly discussed by the Academies in Vienna and Berlin, the Paris Sorbonne and the Budim and Vatican councils. Faced in later years with renewed rumours of epidemic proportion, the Empress Maria Theresia asked her personal physician Gérard Van Swieten to investigate the matter, Louis XV ordered the Duc de Richelieu to find out as much as he could, and the Prussian King Friedrich-Wilhelm I and lesser potentates of Germanic countries instigated similar inquiries; apparently, only King George II did not have questions. He staunchly believed in the existence of vampires.

Various explanations were offered. Some people of the cloth at first accepted the veracity of the alleged phenomena and attempted to find for them the proper theological context. Already Cardinal Prospero Lambertini, who later became Pope Benedict XIV, had very harsh words about 'these priests who gave credit to such stories, in order to encourage simple folk to pay them for exorcism and masses.'[8] Others opted for medicine and common sense. To them many stories were inventions or distortions; certain cases were to be understood as the consequence of food poisoning or the use of opium; in others, it was a matter of corporeal liquids only resembling blood. More systematic approaches included arguments about premature burial, unusually well-preserved corpses under exceptional soil conditions, natural growth of hair and nails after death, diseases, especially the plague, theological confusions due to the conflicts between Roman Catholicism and the Orthodox Churches, heresy, communal superstitions developing under the stress of national disasters and social contradictions. Dom Calmet, while leaning towards a sceptical point of view and duly mentioning all these theories, remained himself strangely undecided:

Mais supposant qu'il y ait quelque réalité dans le fait de ces Apparitions des Vampires, les attribuera-t-on à Dieu, aux Anges, aux Ames de ces Revenans, ou au Démon? ...
Il faut donc demeurer dans le silence sur cet article, puisqu'il n'a pas plû à Dieu de nous révéler, ni quelle est l'étendue du pouvoir du Démon, ni la manière dont ces choses se peuvent faire. ...
Car rien n'est plus mal fondé que ce qu'on dit des Apparitions, des véxations, des troubles causés par les prétendus Vampires & par les Brucolaques. Je ne suis pas surpris que la Sorbonne ait condamné les exécutions sanglantes & violantes, que l'on exerce sur ces sortes de corps morts; mais il est étonnant que les Puissances séculaires & les Magistrats n'emploient pas leur autorité & la sévérité des Loix, pour les réprimer. ...
Dans tout cela je ne vois que ténèbres & difficultés, que je laisse à résoudre à de plus habiles & plus hardis que moi.[9]

Some of the main figures of French Enlightenment found Dom Calmet's hesitations quite ridiculous. Under the word 'Vampire' D.J., i.e. Louis, chevalier de Jaucourt, one of the most assiduous contributors to Diderot's *Encyclopédie ou dictionnaire raisonné des sciences, des arts et des métiers* (v. XVI, 1765, p. 828) commented: 'Le p. Calmet a fait sur ce sujet un ouvrage absurde dont on ne l'auroit pas cru capable, mais qui sert à prouver combien l'esprit humain est porté à la superstition.'[10] And Voltaire's own *Dictionnaire philosophique,* after the publication of the Supplement in the 1765 Varberg edition, contained a jibe under the heading 'Résurrection':

Le profond philosophe dom Calmet trouve dans les vampires une preuve bien concluante. Il a vu de ces vampires qui sortaient des cimetières pour aller sucer le sang des gens endormis; il est clair qu'ils ne pouvaient sucer le sang des vivants, s'ils étaient encore morts: donc ils étaient ressuscités; cela est péremptoire.[11]

Voltaire returned to the same topic in his *Questions sur l'Encyclopédie* (neuvième partie, 1772) in a text under the title 'Vampire,' which was later added to the *Dictionnaire philosophique* by the publishers of the posthumous Kehl edition of the *Œuvres complètes* and since cited as belonging to his 'Dictionary.' The main paragraphs again displayed biting irony and a metaphoric use of the word 'vampire' which was by then common in Western languages:

Quoi! c'est dans notre XVIIIe siècle qu'il y a eu des vampires! ... C'était en Pologne, en Hongrie, en Silésie, en Moravie, en Autriche, en Lorraine, que les morts faisaient cette bonne chère. On n'entendait point parler de vampires à Londres, ni même à Paris. J'avoue que dans ces deux villes il y eut des agisteurs, des traitants, des gens d'affaires, qui sucèrent en plein jour le sang du peuple; mais ils n'étaient point morts, quoique corrompus. Ces suceurs véritables ne demeuraient pas dans des cimetières, mais dans des palais fort agréables. ... les rois ne sont pas, à proprement parler, des vampires. Les vrais vampires sont les moines, qui mangent aux dépens des rois et des peuples.[12]

A parenthesis may be in order here. The first instance of a metaphorical use of 'vampire' probably occurred in a witty article on 'Political Vampires' published in May 1732 by *The Gentleman's Magazine* of London:

This account, of *Vampyres,* you'll observe, comes from the Eastern Part of the World, always remarkable for its *Allegorical Style.* The States of *Hungary* are in subjection to the *Turks* and *Germans,* and govern'd by a pretty hard Hand; which obliges them to couch all their Complaints under *Figures.* This Relation seems to be of the same kind.

These *Vampyres* are said to torment and kill the *Living* by *sucking out all their Blood;* and a *ravenous Minister,* in this part of the World, is compared to a *Leech* or *Bloodsucker,* and carries his Oppressions beyond the Grave, by anticipating the *publick Revenues,* and entailing a Perpetuity of *Taxes,* which must gradually drain the Body Politick of its Blood and Spirits. In like manner, Persons who groan under the Burthens of such a *Minister,* by selling or mortgaging their estates, torment their *unhappy Posterity,* and become *Vampyres* when dead. Paul Arnold, who is call'd a Heyduke, was only a *Ministerial Tool,* because it is said he had kill'd but 4 Persons; whereas, if he had been a *Vampyre* of any Rank, we should probably have heard of his *Ten Thousands.* ... History, especially our own, supplies us with so many Instances of *Vampyres,* in this Sense, that it would fill Volumes to enumerate them. ... Private Persons may be *Vampyres*, or *Blood-Suckers*, i.e. *Sharpers, Usurers, and Stockjobbers, unjust Stewards and the dry Nurses of the Great Estates;* but nothing less than the Power of a *Treasury* can raise up a compleat *Vampyre*.[13]

Throughout the eighteenth century, this usage can be attested in English, French and German letters. Among many examples, in Christoph-Martin Wieland's *Schach Lolo oder das göttliche Recht der Gewalthaber. Eine morgenländische Erzählung* (1778) there is the following: 'Nicht Menschen mehr, *Vampyren* nur erblickt,/ Die an ihm saugen und ihn nagen.'[14] (Five years later, Wieland used the word again, this time referring, albeit metaphorically, to the original meaning — *Clelia und Sinibald oder die Bevölkerung von Lampeduse. Ein Gedicht in zehn Büchern*, 1783.)[15] Oliver Goldsmith invented in his satirical *The Citizen of the World; or Letters from a Chinese Philosopher, Residing in London, to his Friends in the East* a Colonel Leach and a Major Vampire, and he linked certain judges to vampirism: 'a corrupt magistrate may be considered a human hyena, ... [he] sucks blood like a vampyre.'[16] The *Songes d'un hermite* (1770) of Louis Sebastien Mercier contained in the guise of the thirteenth dream 'Les Vampires,' a satirical piece rejecting old wives' superstitions and condemning social bloodsuckers. A similar metaphorical use of the word vampire, linking him to booksellers, is present in Robert Burns' poem 'Thou Nature, partial Nature, I arraign ...'[17]

Earlier in the century, the Swedish botanist Carolus Linnaeus (Carl von Linné) had erroneously applied the term to a harmless South American fruit bat. Later on, Parisian newspapers featured reports about vampire-like monsters which had been observed in South America, for example the *Courrier de l'Europe* in October 1784. And even to-day, any of several Central and South American flying mammals of the genera Desmodus, Diphylla, and Diaemus, which feed on the blood of man and other vertebrates, are known to zoology as vampires or vampire bats.

But let us return to the preternatural vampire. Rousseau as well treated this matter twice, although in passing and in a spirit different from that of Voltaire. In the first draft of his treatise on education, *Emile*, he used this example to make a point about the Church's reliance on miracles, and a little later, in defence of the same *Emile*, he printed an open letter to the Archbishop of Paris (*Jean-Jacques Rousseau, Citoyen de Genève, à Christophe de Beaumont, Archevêque de Paris.* Amsterdam, 1763). In his letter, Rousseau balks at Dom Calmet's apparent attempt to encourage an act of faith by use of factual evidence:

S'il y a dans le monde une histoire attestée, c'est celle des Wampirs. Rien n'y manque; procès verbaux, certificats de Notables, de Chirurgiens, de Curés, de Magistrats. La preuve juridique est des plus complettes. Avec cela, qui est-ce qui croit aux Wampirs? Serons-nous tous damnés pour n'y avoir pas cru?[18]

In written literature, the first use of the vampire motif was made by Heinrich August Ossenfeld, a German poetaster, in his poem 'Mein liebes Mägdelein glaubet ...'; the I proposes to the girl 'Und wenn du sanfte schlumerst ... [ich werde dich] Und als ein Vampir küssen.'[19] These playful lines were deemed by nineteenth-century critics to be very daring ('sehr anstößig' is the expression used by Erich Schmidt in 1887).[20] In oral literature, besides a wide range of typical legends (usually classified today as 'Memorate's), there were a few eighteenth-century versions of two types of the folktale. These are, in Aarne's and Thompson's classification, the Type Nr. 307: The Princess in the Shroud ('Die Prinzessin im Sarge und die Schildwache' in the Grimm manuscript) and especially the popular Type Nr. 365: The Dead Bridegroom Carries off his Bride (sometimes referred to as the 'Fiancée du Mort' or 'Lenore' type). Gogol''s story 'Vij' and Juliusz Słowacki's 'Królewna strzyga,' works of the first half of the nineteenth century, were based on the first type; Gottfried August Bürger gave world prominence to the second in his ballad 'Lenore' (1773). While Bürger did not designate the unquiet bridegroom a vampire but only a revenant and a killer, his probable sources – often studied in great detail since – included tales with explicitly vampiric traits.[21]

Nevertheless, the vampire entered great literature in a work by Goethe, who, as in many other areas, seemed to be able to combine and express the opposing tendencies of the times. A child of the cosmopolitan century, Goethe was the first to express the idea of world literature and to realize it in his own works which then rapidly became 'a spiritual bond among the nations.' Fritz Strich, who used these words in his book *Goethe und die Weltliteratur* (1946), went on to describe the following principal sources of Goethe's inspiration: the liberating impact of Shakespeare's drama, the classical balance provided by Italy, the formative discipline taught by French letters, the example of spirituality given by Spanish theatre, the expanding cultural horizon offered by the Orient, and the democratic spirit embodied by America. But in the context of the vampire-motif, Goethe's important contribution was the ballad 'Die Braut von Korinth,' and its roots are in part elsewhere, in more obscure soil.[22] As often in his writings, Goethe borrowed from Antiquity, but transposed here the subject matter into a startling example of his own shocked fascination with the daemonic. The poet was, needless to say, aware of the darker side of this great ballad; by now well into his classical period, the courtier of Weimar wrote in his diary of Jena under the 4th of June 1797: 'Anfang des Vampyrischen Gedichtes' ('Beginning of the vampiric poem'). In the evening he noted a discussion with Schiller about the 'Romanze' ('romance'), and already under the 5th of June the same diary mentioned 'das Ende des vampyrischen Gedichtes' ('conclusion of the vampiric poem'). This poem, the first in world literature to be based on the vampire motif, was published a year later by Schiller in his *Musenalmanach für das Jahr 1798.*

Directly or indirectly, Goethe took the main plot from the *Antropodemus Plutonicus* (1666) by Johannes Praetorius, who printed a hellenistic tale recorded in the time of the Emperor Hadrian (117-138) by Phlegon of Tralles in his 'Book of Wanders' (*Peri thaumiōn kai makrobiōn*) and later copied by Proclus (412-485). The source tells a simple ghost story about a lecherous wench and it is only Goethe who combined this anecdote with two other topics, one being the problem of conflicting religious beliefs and the other the theme of the dead's inalienable right to love. This right had been mentioned by him in the *Sturm und Drang* ballad 'Es war ein Buhle ...' ('There was a dallier ...') of 1774, better known under the later title 'Der untreue Knabe' ('The faithless youth'). In 'The Bride of Corinth' he added to this right of the dead traits derived from Serbian and Dalmatian beliefs about vampires. He knew the Serbian superstition from Dom Calmet's notorious *Dissertations,* a source very rarely mentioned in Goethe scholarship although the French historian already connected Phlegon's story with vampirism;[23] information about the Dalmatian not quite dead was obtained from the travelogue of the abbate Alberto Fortis, i.e. his *Viaggio in Dalmazia.* The rele vant chapter of this book on the customs of the Morlacchi ('De' costumi de' Morlacchi') had provided him twenty years earlier with the Italian text of the Serbian ballad 'Hasan-aginica,' which as 'Klaggesang der edlen Frauen des Asan Aga' became in Goethe's translation an internationally accepted example of folk poetry. The rather brief passage in Fortis about the existence of vampires contains also one of the oldest references to Transylvania.[24] Regardless of Slavic additions, Goethe preserved the Greek setting, holding fast to the conviction derived during his trip to Italy that Antiquity is the ideal basis of European civilization.

The content of the ballad is well known. A pagan Athenian youth comes to Corinth to claim a promised bride. Since the betrothal by the parents, the girl's family has embraced the new Christian creed and sent the daughter to a nunnery, where she has died of sorrow. Ignoring these events and arriving late at night, the youth is led to a guest room. In the middle of the night, the maiden comes to his room and couch, to partake with him in the gifts of Ceres and Bacchus. The true nature of the girl is suggested by stages: her pallid hand and pale face, the request of a ringlet of his hair, her eager drinking of draughts of dark purple wine, her shuddering dread of the wheaten bread, the ice cold limbs and the breast without heartbeat. The impassioned youth insists: 'Hoffe doch bei mir noch zu erwarmen, / Wärst du selbst mir aus dem Grab gesandt!' ('I could warm thee well with my desire, / Wert though even from the grave ascended!').[25] The poem reaches the tragic turning-point when the mother, hearing love's mysterious stammerings and fiery protestations, enters the chamber expecting to find a shameless minion, but espies her own deceased daughter. The maiden 'hebet ... lang and langsam sich im Bett empor' ('long and slowly rises from the bed') and arraigns her mother's ascetic religion, because 'die Erde kühlt die Liebe nicht' ('love must still burn on, though wrapped in clay'); she proclaims at last her metamorphosis:

Aus dem Grabe werd' ich ausgetrieben,
Noch zu suchen das vermißte Gut,
Noch den schon verlornen Mann zu lieben
Und zu saugen seines Herzens Blut.
Ist's um den geschehn,
Muß nach andern gehn,
Und das junge Volk erliegt der Wut.[26]

She will rest in peace only if her corpse is burned together with that of the youth, because he, too, must soon decay, his black hair turn gray and his body die.

It is instructive to compare the last stanza of the ballad ('Eilen wir den alten Göttern zu' – 'To the ancient gods aloft we soar') with the ending of Bürger's 'Lenore.' Baroque images of man's mortal coil are followed in 'Lenore' by the howling of ghosts; crying out in their dance of death, they proclaim the victory of unyielding destiny and confirm *in extremis* the mother's Lutheran credo:

Geduld! Geduld! Wenn's Herz auch bricht!
Mit Gottes Allmacht hadre nicht!
Des Leibes bist du ledig;
Gott sei der Seele gnädig![27]

On the contrary, in 'The Bride of Corinth' tensions are not dissolved in religious belief and no solace is offered. The last words are uttered by the girl, who proclaims her defiant wish to soar to the *ancient gods* on the sparks of the funeral pyre. The ballad in this way sharpens the contrast between the hellenistic, pagan enjoyment of the senses and Christian self-denial, and permits a reading which rejects Christianity's fear of the body's instincts and appetites. (Whoever wishes to read more of Goethe's extreme attacks against Christianity and even against the person of Christ, may consult his surpressed *Venezianische Epigramme*, especially those taken in the twentieth century from sealed manuscripts, published in the last supplements to the monumental 'Sophien Ausgabe' of Weimar, and – to the best of my knowledge – never reprinted since.)[28]

Goethe's poem provoked a minor sensation. Of those who were close to the poet, only Wilhelm von Humboldt, the philologist and diplomat, showed sincere enthusiasm. Herder, at that time head ('General-superintendant') of the Lutheran church in Weimar and altogether cool towards his erstwhile close friend, rejected the poem as incompatible with his own Christian and spiritualistic convictions, and spoke acrimoniously of a priapic, revolting and disgusting topic.[29] Even Madame de Staël condemned, in *De l'Allemagne* (1810), the 'volupté funèbre' ('deadly voluptuousness') of the poem,[30] and French translators and literary critics shied away from the 'paganisme de son sujet' ('paganism of its subject').[31] They simply eliminated the offending passages, which had not fared better in Madame de Staël's detailed summary. To illustrate one characteristic nineteenth-century voice, the historian Jules Michelet, in his *La Sorcière* (1862), which is, incidentally, a book against the mediaeval church and in favour of nature, regards Goethe's treatment of the antique story as a 'pollution':

Goethe, si noble dans sa forme, ne l'est pas autant d'esprit. Il gâte la merveilleuse histoire, souille le grec d'une horrible idée slave. Au moment où l'on pleure, il fait de la fille un vampire. Elle vient parce qu'elle a soif du sang, pour sucer le sang de son cœur. Et il lui fait dire froidement cette chose immonde: "Lui fini, *je passerai à d'autres;* la jeune race succombera à ma fureur."[32]

True enough, the Romantics reacted otherwise and Friedrich Schlegel declared, in one of the programmatic fragments of 1798, that this ballad (he called it a romance) was an epitome of the genre and that it 'is epoch-making in the history of poetry' ('Epoche in der Geschichte der Poesie macht').[33] Indeed, in the twentieth century critics have also praised the immense poetic strength of the work, but have continued to be perturbed by the meaning. Three examples may stand for many. Max Kommerell in his famous *Gedanken über Gedichte* steadfastly rejected anti-Christian interpretations:

eine heftige, schwül gespannte Liebesstimmung, die ins Gespenstische umschlägt, schließlich die feierlich beschwörende Verrufung der christlichen Frevel an Liebe, Leib und Blut aus dem Munde der Wiedergängerin; ... Die letzten Strophen rechnen unter die größten sprachlichen Taten dieses Dichters. Altertümlich ist also nur das Motiv, das weniger antik als slavisch ist, ein echtes Balladenmotiv, nur daß es hier eine weltgeschichtliche Tragweite bekommt. ... Dennoch ist es nicht erlaubt, das Gedicht als antichristliches Manifest aufzufassen. Es ist Ballade. ... Es gibt in Goethes Werk mehrere Ordnungen und mehrere Gerichtsbarkeiten. Und nur alle zusammen sind Goethe selbst.[34]

Emil Staiger, in the Artemis memorial edition reacted to this ballad and to 'Der Gott und die Bajadere' as to 'unheimliche fremde Gäste' ('uncanny alien guests'):

Wie fremd bleibt ... der innige Bund von Liebe und Tod, wie fremd zumal die Vampyrsage mit ihrer verführerisch grausigen Stimmung! Gleich einem ungeheuren Traum, zu dem der Erwachte sich kaum zu bekennen wagt, muß das Gedicht sich von der Seele Goethes losgelöst haben. Wir hören denn auch, daß es als Geheimnis jahrzehntelang in ihm verborgen lag, bevor es in solcher Vollendung, in solcher dämonischen Stille, in diesen Strophen von evokativer Kraft als Kunstwerk gestaltet werden konnte. Einige Freunde, darunter bedeutende Kenner des Dichters, waren bestürzt.[35]

And one of the most recent interpreters, the academician Sergej Ivanovič Sobolevskij totally neglects the vampire element in his study 'Goethe's "Bride of Corinth" and its Antique Source' and then by necessity finds certain parts to be 'enigmatic' and 'unexplainable.'[36]

Goethe mentioned the vampire later, in passing, in the Second *Faust*,[37] and praised John Polidori's *The Vampyre*, which he, with many contemporaries, mistook for a work by Lord Byron.[38] Nevertheless, the aging court counsellor spoke often enough against 'morbid exaggerations,' Kleist's 'perversions,' the 'sick' and 'pathologic' romanticism, castigating the 'excesses' of that 'ultra-romantic' tendency which, in his words, swamps literature with infamies and revolting ugliness, and choses its heroes among thieves and galley slaves, vampires and monsters. Victor Hugo's *Notre-Dame de Paris* merited the distinction of being the most disgusting novel ever written.[39] Still, Goethe had himself given an early example of such ambiguities. He provided the very first, and for this reason the most disturbing, image of one of man's existential dualities: a poetic picture of that impassioned love which is also an obsession with death, of that lustful marriage of Eros and Thanatos which prefigures the voluptuous mating of Freud's sex and death instincts. Goethe's vampirism was, as a result, not only a forerunner of the romantic agony depicted by Mario Praz in his inquiry into the beauty of the Medusa, the metamorphoses of Satan, the shadow of the divine Marquis, and *la belle dame sans merci.*[40] It was also a milestone on the road to romantic exploitations of this motif in works by Southey, Coleridge, Lamb, Moore, Polidori, Shelley, Keats, Poe, Tieck, Brentano, Hoffmann, Nodier, Mérimée, Dumas, Baudelaire, Potocki, Mickiewicz, Gogol', Aleksej Tolstoj, and others. 'The Bride of Corinth' is likewise the direct impetus of such works as Théophile Gautier's novella 'La Morte amoureuse' (1836) and his poem 'Les Taches jaunes' (1844).[41] It is tempting – and comparatists succumb easily to such blandishments – to perceive an uninterrupted red thread leading from this ballad to the peculiar concept about the relationship of love, beauty, and death expressed by August von Platen in the brief fragment 'Tristan' (1825):

Wer die Schönheit angeschaut mit Augen,
Ist dem Tode schon anheimgegeben,
Wird für keinen Dienst der Erde taugen,
Und doch wird er vor dem Tode beben,
Wer die Schönheit angeschaut mit Augen. ...[42]

One finds here a concept which Richard Wagner and Thomas Mann later used so skillfully as an element of their large embroidered tapestries of music, poetry, and fiction. Love and death, love stronger than death, death which kills lovers – these are motifs which existed before Goethe and which can be traced back to Greek mythology. But his ballad is the first poetically perfect rendering of the violent erotic *Liebestod.*[43]

Let us return to the opening proposition. Throughout the century – as shown by Georges Gusdorf and Roland Mortier among many contemporary scholars[44] – the light of reason had to contend with that of sentiment, passion, enthusiasm, and genius, the striving for order and culture with that for freedom and nature. If Voltaire opted for scepticism and irony, Rousseau for the voice of intuition and emotion, Diderot and Goethe demonstrated the coincidence or alternation of both motivations within one life. Roland Mortier is justified, therefore, in rejecting the more traditional scheme of an eighteenth century characterised by a succession of three distinct phases: critical reason, preromantic sensibility, and romanticism, and in concluding that the century was founded on the dialectic between two main tendencies, that of the rational which dominates towards the beginning, and that of the irrational which becomes increasingly prominent towards the end.[45]

It is useful to remind oneself that vampiromania, so popular in paraliterature since Polidori and Bram Stoker, and a favourite of modern moviegoers and T.V. watchers, had already titillated the reading classes of the self-consciously civilized and enlightened parts of eighteenth-century Europe. And there is a more serious side to the delightful shivers provided by this motif. Borrowed from the darkest collective beliefs of the less progressive oral cultures of Eastern, Central, and South-Eastern Europe, the idea of the vampire met and merged with the deep shadows stretching from the dim, ancient past and reaching up from within the unfathomable depths of the Western psyche, revealing its powers of horror, its own abjection.[46] Even Goethe, ever endeavouring to overcome a certain 'daemonic side' ('das Dämonische') in the world and in himself,[47] nourished in his subconsciousness for long years the archetypal fear of the dead, the primordial dread of the woman, and for a creative moment succumbed to both.

MILAN V. DIMIĆ
University of Alberta

Notes

1 For these contrasts see, among many other studies, Auguste Viatte, *Les Sources occultes du romantisme, illuminisme — théosophie, 1770-1820.* Bibliothèque de la Revue de Littérature Comparée, 46-47, 2 vols. (Paris: H. Champion, 1928); Georges Gusdorf, *Les Sciences humaines et la pensée occidentale. VII: Naissance de la conscience romantique au siècle des lumières* (Paris: Payot, 1976). For the rise of black humour in the eighteenth century see Patrick O'Neill, 'The Comedy of Entropy: The Contexts of Black Humour,' *Canadian Review of Comparative Literature/ Revue Canadienne de Littérature Comparée,* 10 (1983), esp. 149-156.

2 R.D. Scott, *The Holy and the Daemonic from Sir Thomas Browne to William Blake* (Princeton, NJ: Princeton University Press, 1982). See also the same author's, 'The Witch of Endor and the Gadarene Swine: The Debate over Witchcraft and Daemonianism in the Seventeenth and Eighteenth Centuries,' *Essays in Literature,* 9 (1982), 143-154.

3 Some of the more useful general studies and anthologies of vampirism are: Stefan Hock, *Die Vampyrsagen und ihre Verwertung in der deutschen Litteratur,* Forschungen zur neueren Litteraturgeschichte, 17 (Berlin: A Duncker, 1900; rpt. Hildesheim: Gerstenberg, 1977); Ernst Havekost, *Die Vampyrsage in England* (Halle a.S.: privately printed, 1914); Dudley Wright, *Vampires and Vampirism,* 2nd rev. ed. (London: W. Rider, 1924; rpt. as *The Book of Vampires,* New York: Causeway Books, 1973); Montague Summers, *The Vampire: His Kith and Kin* (London: K. Paul, 1928; rpt. New Hyde Park, NY: University Books, 1960); same, *The Vampire in Europe* (London: K. Paul, 1929; rpt. New Hyde Park, NY: University Books, 1968); Tihomir Đorđević, *Vampir i druga bića u našem narodnom verovanju i predanju,* ed. V.S. Radovanović, Srpska Akademija Nauka, Od. društvenih nauka, Srpski etnografski zbornik, 66, drugo od., Život i običaji narod-

ni, 30 (Beograd: S.A.N., 1953); Ornella Volta and Valerio Riva, eds., *I vampiri tra noi* (Milano: G. Feltrinelli, 1960; French adaptation as *Histoires de vampires,* Paris: R. Laffont, 1961); Ornella Volta, *Le Vampire: la mort, le sang, la peur,* Bibl. internationale d'Erotologie, 8 (Paris: Pauvert, 1962); Dieter Sturm and Klaus Völker, eds., *Von denen Vampiren oder Menschensaugern: Dichtungen und Dokumente* (München: C. Hauser, 1968); Raymond T. McNally and Radu Florescu, *In Search of Dracula: A True History of Dracula and the Vampire Legends* (Greenwich, CT: New York Graphic Society, 1972); Aribert Schroeder, *Vampirismus: Seine Entwicklung vom Thema zum Motiv,* Studien zur Anglistik, 8 (Frankfurt a.M.: Akademische Verlagsgesellschaft, 1973); Jan L. Perkowski, *Vampires of the Slavs* (Cambridge, MA: Slavica Publishers, 1976): Christopher Frayling, ed., *The Vampire: Lord Ruthven to Count Dracula* (London: V. Gollancz, 1978); Harry A. Senn, *Were-Wolf and Vampire in Romania,* East European Monographs, 99 (New York: Columbia University Press, 1982). These studies notwithstanding, all original sources mentioned in the text have been used *de visu.*

4 Paul Hazard, *La Crise de la conscience européenne, 1680-1715* (Paris: Boivin, 1935).

5 For a Slavic origin see among the old sources J.J. Hanus[h], 'Die Vampyre,' *Zeitschrift für Deutsche Mythologie und Sittenkunde,* 4 (1859), 198-201, esp. 200-201; among the newer Summers, *The Vampire,* pp. 18-20; Milan Budimir, 'Vampirizam u evropskoj književnosti,' *Anali Filološkog fakulteta,* 6 (1966), 268-73, esp. 271-72; Schroeder, pp. 13-28; and most standard etymological dictionaries of European languages. For the first instance of the often cited 'Turkish' theory, see Franz Miklosich, *Etymologisches Wörterbuch der slavischen Sprachen* (Wien: W. Braumüller, 1886), pp. 374-75.

6 For the Austrian reports see part II, ch. xi, pp. 278ff., 281ff.

7 The report is printed in most sources mentioned in note 3 and also in Dom Calmet's book (see notes 6 and 9). With small corrections, I am following the English translation in Frayling, pp. 28-31.

8 *De Servorum Dei Beatificatione et de Beatum Canonizatione* (Romae: Salomoni, 1749 and 1756), liber IV, pars I, caput xxi, pp. 323-24 (my translation).

9 Dom Augustin Calmet, *Traité sur les apparitions des esprits, et sur les vampires, ou les revenans de Hongrie, de Moravie, & c.,* nouvelle éd. (Paris: Debure l'ainé, 1751), II, 257, 259, 305, and 307 respectively. (In Frayling's translation, pp. 34f.: 'Supposing there be any reality in the fact of these apparitions of vampires, shall they be attributed to God, to angels, to the spirits of these ghosts, or to the devil? ... We must remain silent on this matter, since it has not pleased God to reveal to us either the extent of the demon's power, or the way in which such things may come to pass ... The stories told of these apparitions, and all the distress caused by these supposed vampires, are totally without solid proof. I am not surprised that the Sorbonne has condemned the bloody and violent retribution wrought on these corpses; but it is astonishing that the magistrates and secular bodies have not employed their authority and legal force to put an end to it ... This is a mysterious and difficult matter, and I leave bolder and more proficient minds to resolve it ...')

10 'Father Calmet has written an absurd book on this subject, a book one would not have thought him capable of writing. It only goes to show how far the human mind is prone to superstition.' (Translation in Frayling, p. 35.)

11 Voltaire, *Œuvres complètes*, nouvelle éd. [L. Moland] conforme pour le texte à l'éd. de Beuchot (Paris: Garnier Frères, 1879; rpt. Nendeln/Lichtenstein, 1967), XX, 367. ('The profound philosopher Father Calmet found in vampires a conclusive proof. He has seen vampires come out from graveyards to go and suck the blood of sleeping people; it is clear that they could not suck the blood of the living if they were still dead: therefore, they were resurrected; this is imperative.' My translation.)

12 Voltaire, pp. 547-51, the quoted sentences are on p. 547, 548, and 550. ('What! Vampires in our Eighteenth Century? Yes ... in Poland, Hungary, Silesia, Moravia, Austria and Lorraine – there was no talk of vampires in London, or even Paris. I admit that in these two cities there were speculators, tax officials and businessmen who sucked the blood of the people in broad daylight, but they were not dead (although they were corrupted enough). These true bloodsuckers did not live in cemeteries: they preferred beautiful palaces. ... Kings are not, properly speaking, vampires. The true vampires are the churchmen who eat at the expense of both the king *and* the people.' Translation in Frayling, p. 37.)

13 Quoted in Frayling, pp. 33-34.

14 Wieland, *Gesammelte Schriften: Dichtungen I*, ed. W. Kurrelmeyer, I. Abt. (Berlin: Preußische Akademie der Wissenschaften, 1935), XII, 332, vv. 180-81. ('Shah Lolo or the Divine Right of Potentates: An Oriental Tale: Not People, but only vampires he can see, who suck him and devour.' My translation.)

15 *Dichtungen II*, XIII, 338, vv. 240-43. ('Clelia or Sinibald or the Inhabitants of Lampeduse: A Poem in Ten Books')

16 Goldsmith, *Collected Works*, ed. Arthur Friedman (Oxford: Oxford University Press, 1966), II, 206 and 329 (quote).

17 Burns, *Poems Chiefly in the Scottish Dialect*, 2nd ed. (Edinburgh: T. Cadell and W. Creech, 1793), II, 183.

18 Rousseau, *Œuvres complètes*, éd. Bernard Gagnabin et Marcel Raymond, Bibl. de la Pléiade, 208 (Paris: Gallimard, 1969), IV, 987. ('If there is in this world one story that has been attested, it is that of Vampires. No evidence is lacking: the testimonies on oath, certificates signed by Notables, Surgeons, Priests, Magistrates. The proof *in law* is thoroughly convincing. Yet with all this, who actually *believes* in Vampires? Will we all be damned for not believing in them?' Translation in Frayling, p. 37.) The early draft of the MS. Favre of *Emile* is published in *La 'Profession de foi du vicaire Savoyard' de Jean-Jacques Rousseau*, éd. Pierre-Maurice Masson, Collectanea friburgensis, NS. 16 (Fribourg: O. Gschwend, 1914), cf. p. 331, n. 3.

19 'Der Vampir,' *Der Naturforscher*, 48 (Leipzig, 25. 5. 1748), 380-81. Rpt. in Hock, p. 45, but not in Ossenfelder, *Oden und Lieder*, 4 Theile in 1 Band (Dresden und Leipzig: J.W. Harpetern, 1753). ('My dear sweetheart believes ... And when you slumberst peacefully ... [I Will] kiss you like a vampire.' My translation.)

20 Erich Schmidt, 'Ossenfelder, Heinrich August,' in *Allgemeine Deutsche Biographie*, 24 (1887), 498.

21 See i.a. Wilhelm Wackernagel (and Heinrich Hoffmann), 'Zur Erklärung und Beurtheilung von Bürgers Lenore,' *Altdeutsche Blätter*, 1 (1836), 174-204; W. Wollner, 'Der Lenorenstoff in der slavischen Volkspoesie,' *Archiv für slavische Philologie*, 6 (1882), 239-69; Bogomil Krek, 'Ein Beitrag zur Literatur des Lenorenstoffes,' *Archiv*

für slavische Philologie, 10 (1887), 356-59; Karl Krumbacher, 'Ein Problem der vergleichenden Sagenkunde und Litteraturgeschichte (Die Lenorensage),' *Zeitschrift für vergleichende Litteraturgeschichte*, 1 (1887), 214-20; Nikola Andrić, 'Bajka o Lenori,' *Vienac*, 25 (1893), 513-18, 528-30; Heinrich von Wlislocki, 'Zur Lenorensage,' *Zeitschrift für vergleichende Litteraturgeschichte*, NS. 11 (1897), 467-74; [Carl] Mengis, 'Lenore,' *Handwörterbuch des deutschen Aberglaubens*, 5 (1932/33), 1209-1211; Will-Erich Peuckert, *Lenore*, FFC, 158 (Helsinki: Suomalainen Tiedeakatemia, 1955); Gh. Vrabie, 'Călătoria fratelui mort sau motivul Lenore in folclorul sud-est European.' *Limba şi literature*, 3 (Bucureşti, 1957), 257-94 (revised in G.V., *Balada populara română*, Bucureşti, 1966, 108-144); Felicia Marinca, 'O variantà a ballada călătoria fratelui mort (motivul Lenore),' *Limba şi literatura*, 6 (1962), 387-95; Lutz Röhrich (and Rolf Wilhelm Brednich), *Deutsche Volkslieder I: Erzählende Lieder* (Düsseldorf: Schwann, 1965), pp. 65ff., No. 8; Jerzy Śliziński, 'Zum Lenorenmotiv in den westslawischen Literaturen,' *Wissenschaftliche Zeitschrift der Humboldt-Universität zu Berlin, Gesellschaftlich- und Sprachwissenschaftliche Reihe*, 16 (1967), 749-53; Dagmar Burkhart, 'Nachträge zum Lenoren-Motiv auf dem Balkan,' in *Actes du Premier Congrès international des études balkaniques et sud-est européennes* (Sofija, 1971), II, 863-68; Maria Ioannidu-Barbarigou, '"La Chanson du Frère Mort": Les Variantes du Pont,' *Actes du IIe Congrès International des Etudes du Sud-est Européen (1970), V: Linguistique et Littérature* (Athènes: Assoc. Internationale des Etudes du Sud-est Européen, 1978), 731-40 (with an overview of the older critical literature in Greek). For literary interpretations see, besides Wackernagel, esp. Erich Schmidt, 'Bürgers 'Lenore,'' in *Charakteristiken I* (Berlin: Weidmann, 1886), pp. 199-244; Friedrich Gundolf, 'Bürgers 'Lenore' als Volkslied,' [1930] rpt. in F.G., *Beiträge zur Literatur- und Geistesgeschichte*, Eds. V.A. Schmitz and F. Martini, Veröffentlichungen der Deutschen Akademie für Sprache und Dichtung, 54 (Heidelberg: Schneider, 1980), 277-98; Albrecht Schöne, 'Bürgers Lenore,' *Deutsche Vierteljahrsschrift für Literaturwissenschaft und Geistesgeschichte*, 28 (1954), 324-44; Lore Kaim, 'Bürgers Lenore,' *Weimarer Beiträge*, 2 (1956), 32-68; Hermann Pongs, *Das Bild in der Dichtung III: Der symbolische Kosmos der Dichtung* (Marburg: N.G. Elwert, 1969), pp. 71-83; Elizabeth Chadwick, 'Poetry as Literate Response: G.A. Bürger and the Ballad Audience in Germany and England 1773-1800,' Diss. Yale 1976-77 (*DA*, 37, 4382A); Louis Vax, *Les Chefs-d'œuvre de la littérature fantastique* (Paris: P.U.F., 1979), pp. 172-74.

22 'Die Braut von Korinth,' in Goethe, *Werke*, Hamburger Ausgabe, ed. Erich Trunz, 12th ed. (München: C.H. Beck, 1981), I, 268-73 (text), 659-64 (commentary). The main critical references to the ballad are: Hock, pp. 66-72; Albert Leitzmann, *Die Quellen von Schillers und Goethes Balladen*, Kleine Texte für Theologische und Philologische Vorlesungen, 73 (Bonn: A. Marcus and E. Weber, 1911; 2nd. ed. 1923), esp. pp. 34-37; Georg Mayer, 'Die Braut von Korinth – eine visionäre Ballade,' *Preußische Jahrbücher*, 181-82 (1920), 30-41; Ludwig Heilbrunn, *Die Braut von Korinth: Vortrag*, Privatdruck der Frankfurter Gesellschaft der Goethe Freunde, 8 (Frankfurt a.M.: Frankfurter Gesellschaft der Goethe Freunde, 1926), 18p.; Paul Ludwig Kämpchen, *Die numinose Ballade: Versuch einer Typologie der Ballade*, Mnemosyne, 4 (Bonn: Röhrscheid, 1930), pp. 10-12, 23-24; Max Kommerell, *Gedanken über Gedichte* (Frankfurt a.M.: V. Klostermann, 1943; 2nd. ed. 1956), pp. 361-64; James Boyd, *Notes to Goethe's Poems II (1786-1832)* (Oxford: B. Blackwell, 1949; rpt. 1958), pp. 82-93; Emil Staiger, *Goethe II: 1786-1814* (Zürich: Atlantis, 1956), pp. 303, 307-15, 346; Albert Zastrau, 'Ballade: Die Braut von Cor-

inth,' in Zastrau, ed., *Goethe Handbuch: Goethe, seine Welt und Zeit in Werk und Wirkung,* 2nd ed. (Stuttgart: J.B. Metzler, 1961), I, 643-664; Vax, pp. 174-78.

23 The conventional wisdom of Goethe-studies is expressed for instance by Erich Schmidt, 'Quellen Goethescher Balladen. a) Zur "Braut von Korinth," ' *Goethe-Jahrbuch,* 9 (1888), 229-34; Eduard von der Hellen, ed., *Goethes Sämtliche Werke: Jubiläums-Ausgabe* (Stuttgart-Berlin: J.G. Cotta, 1912), XL, 347; Leitzmann, p. 34; Mayer, p. 31; Hermann Baumgart, *Goethes lyrische Dichtung in ihrer Entwicklung und Bedeutung* (Heidelberg: Carl Winter, 1933), II, 57; Boyd, II, 85ff. (with reservations); Trunz, p. 663. Exceptions are Hock, pp. 67f., who believes in a contamination of Calmet with Fortis, and Zastrau, who repeatedly signals the general importance of Calmet for Goethe's interest in vampires (*Goethe Handbuch,* 2nd ed., I, 650, 653 – s.v. Ballade, 1535 – s.v. Calmet). For Staiger, the exact source is unknown (Staiger, II, 308).

In any case, Goethe denied the direct use of Phlegon of Tralles. The Chancellor F. von Müller notes in Weimar, under 6 June, 18245: 'Webers Vorlesung über die Braut von Korinth gefiel ihm. Doch fügte er hinzu, habe ich nicht aus Phlegons von Tralles (eines Freigelassenen Kaiser Hadrians) Traktat von wunderlingen Dingen, sondern wo anders her das Sujet genommen, es aber meist vom Stoffartigen entkleidted. Philinnion hieß die Braut.' ('Weber's lecture about the Bride of Corinth had pleased him. Nevertheless he added, I have not taken the theme from the Treatise about Wonders by Phlegon of Tralles [a freedman of the Emperor Hadrian], but from somewhere else, and have purified most elements of the subject matter. The bride's name was Philinnion.' My translation.) Flodoard Frhr. v. Biedermann, *Goethes Gespräche: Gesamtausgabe,* 2nd ed. (Leipzig: F.W. v. Biedermann, 1909-1911), n° 2266, III, 119. For probable family 'table talk' about vampires during Goethe's childhood and youth cf. Franz Götting, 'Die Bibliothek von Goethes Vater,' *Nassauische Annalen,* 64 (1953), 23-69, esp. p. 39.

Dom Calmet juxtaposes this anecdote about 'Philinnium et Machates' with stories about vampires in all versions of his book: Augustin Calmet, *Dissertations sur les apparitions des anges, des démons & des esprits et sur les revenans et vampires ...* (Paris: de Bure l'ainé, 1746), pt. II, ch. v, pp. 265-68; also the same author's, *Traité...* (1751), II, 22-25; etc. The French versions speak of similar stories ('une Histoire qui a quelque rapport à celle-ci'), but the German translations identify the phenomenon with 'Vampiren.' Cf. the same author's, *Gelehrte Verhandlung der Materi von Erscheinungen der Geisteren, und denen Vampiren in Ungarn...,* 3rd ed. (Augspurg: Matthäus Rieger, 1757), 2. Theil, V. Capitul: Auferstehung oder Erscheinung einer ledigen Weibs-Person, welche seit einigen Monaten verstorben ware, II, 15-21, which is immediately followed, on p. 21, by: VI. Capitul: Ein Weib wird lebendig aus dem Grab gezogen. In einem gantz neuen Buch wird eine Geschicht gelesen, welche dame, so man von denen Vampiren erzaehlt, einiger massen aehnlich ist. The story is then compared to that about Machates and Philinnium (p. 22). The 'Hungarian' vampires follow a few pages later (pp. 29ff., 31ff.). The earlier German editions are from 1751, 1752, and perhaps 1755 (cf. *Gesamtverzeichnis des deutschsprachigen Schrifttums /G.V./ 1700-1910,* v. 23, München-New York-Paris, 1980).

24 Abbate Alberto Fortis, *Viaggio in Dalmazia,* 2 vols. (Venezia: A. Milocco, 1774), I, 64 (about vampires), I, 98 ('Asan Aga'). Goethe used the translation by Werthes, *Sitten der Morlacken aus dem Italiänischen übersetzt* (Bern: Typographische *Gesellschaft,* 1775), p. 33 (vampire), p. 90 'Klaggesang'); cf. *Reise zu den*

Morlacken, 2 vols. (Bern: Typographische Gesellschaft, 1776), pp. 95 and 152, respectively.

25 All quotations from the poem follow the edition by Trunz. English translations are provided by W.E. Aytoun and Theodore Martin in Thomas Mann, ed., *The Permanent Goethe,* The Permanent Library Series (New York: The Dial Press, 1948), pp. 353-58.

26 'Nightly from my narrow chamber driven,/ Come I to fulfil my destined part,/ Him to seek to whom my troth was given,/ And to draw the life-blood from his heart./ He hath served my will;/ More I yet must kill,/ For another prey I now depart.'

27 Gottfried August Bürger, *Gedichte,* ed. August Sauer, Kürschner's Deutsche National-Litteratur, 78 (Berlin and Stuttgart: W. Spemann, s.d. [1884]), p. 174. These lines became in Walter Scott's 'William and Helen' (1795), his so-called 'imitation from the German of Bürger,' the following stanza: 'E'en when the heart's with anguish cleft,/ Revere the doom of Heaven!/ Her soul is from her body reft:/ Her spirit be forgiven!' Scott, *Poetical Works: With the Author's Introduction and Notes,* ed. J. Logie Robertson (London: Oxford University Press, 1904; rpt. 1967), pp. 630-34, quotation on p. 634.

28 See *Goethes Werke,* Herausgegeben im Auftrage der Großherzogin Sophie von Sachsen-Weimar, 143 vols., Abt. I: Werke, 55 vols. (Weimar: H. Böhlau, 1887-1919), Abt. I, vol. 53, pp. 8-18, 347-51, 430 (No. 158).

29 Herder is quoted by Boyd, II, 84-85.

30 Germaine Necker, Baronne de Staël-Holstein, *De l'Allemagne,* éd. Comtesse Jean de Pange [and S. Balayé], 5 vols., Les Grands Ecrivains de la France (Paris: 1958-60), II, 202-209, quotation on p. 207.

31 Voyslav M. Yovanovitch, *'La Guzla' de Prosper Mérimée: Etude d'histoire romantique* (Paris: Hachette, 1911), pp. 318-19.

32 *La Sorcière* (Paris: Club Français du Livre, 1959), p. 333. ('Goethe, so noble in form, does not reach the same perfection in meaning. He spoils the marvellous story, pollutes the Greek with a horrible Slavic notion. The very moment we are ready to shed tears, he transforms the girl into a vampire. She comes because she thirsts for blood and wants to suck his heart-blood. And Goethe lets her coldly state this filth: "After finishing him off, *I am going to the others,* the young generation will succumb to my fury." ' My translation)

33 'Wie die Novelle in jedem Punkt ihres Seins und ihres Werdens neu und frappant sein muß, so sollte vielleicht das poetische Märchen und vorzüglich die Romanze unendlich bizarr sein; ... Es gibt eine Bizarrerie der Begeisterung, die sich mit der höchsten Bildung und Freiheit verträgt, und das Tragische nicht bloß verstärkt, sondern verschönert und gleichsam vergöttlicht; wie in Goethes "Braut von Korinth," die Epoche in der Geschichte der Poesie macht. Das Rührende darin ist zerreißend und doch verführerisch lockend. Einige Stellen könnte man fast bürlesk nennen, und eben in diesen erscheint das Schreckliche zermalmend groß.' Friedrich Schlegel, 'Fragmente,' *Athenäum: Eine Zeitschrift von August Wilhelm Schlegel und Friedrich Schlegel,* 1. Bd., 2. Stück (Berlin, 1798), No. 429; rpt. in *Goethe im Urteil seiner Kritiker,* ed. Karl Robert Mandelkow (München: C.H. Beck, 1975), I, 155. ('As the novella has to be new and astonishing in every moment of its being

and becoming, so the poetic folk-tale and especially the ballad probably should be infinitely strange; ... There is a bizarrerie of enthusiasm which accords well with the highest refinement and freedom; this bizarrerie not only strengthens the tragic, but also embellishes and almost apotheosizes it. This is the case with Goethe's "Bride of Corinth," which is epoch-making in the history of poetry. Its pathos is lacerating, but still seductively attractive. A few passages could be called almost farcical, but in the very same ones appears a horror of annihilating proportions.' My translation)

34 Kommerell, p. 362 and 364, respectively. ('... a passionate, overheated and tense mood of love, turning into the ghastly; at the end the solemn incantantory curse ... pronounced by the revenant ... The last stanzas belong among the greatest rhetorical achievements of the poet. Old is therefore only the motif, which is less from Antiquity and more Slavic, a genuine balladic motif, but enlarged to proportions of world history ... Nevertheless, it is not permissible to understand the poem as an anti-Christian manifesto. It is a ballad ... There are in Goethe's works several orders and several moral judgements. And only all taken together represent Goethe himself.' My translation)

35 Goethe, *Gedenkausgabe der Werke, Briefe und Gespräche*, ed. Ernst Beutler, I: *Gedichte*, ed. Emil Staiger (Zürich, Artemis, 1950), p. 741. ('How strange remains ... the intimate link between love and death, how strange especially the vampire legend with its seductively horrible mood! The poem must have arisen from Goethe's soul as a monstruous dream, almost unacceptable after waking up. And we know that it slumbered in him for decades as a secret, before he could shape it with such perfect artistry, in such daemonic silence, into these stanzas of extreme evocative power. A few of his friends, among them important connoisseurs of his work, were dumfounded.' My translation)

36 Sergej Ivanovič Sobolevskij, ' "Korinfskaja nevesta" Gëte i eë drevnij original,' *Izvestia Akademii Nauk SSSR, Otdelenie literatury i jazyka*, 19 (Moskva, 1960), No. 4, 284-300, esp. p. 300.

37 *Faust II*, act I, prose text after v. 5298.

38 In a conversation with F. v. Müller, Weimar, Febr. 25, 1820 (No. 1902 in *Goethes Gespräche*, 2nd ed. Biedermann; No. 4756 in *Goethes Gespräche*, ed. Wolfgang Herwig, 4 vols. [Zürich-Stuttgart: Artemis, 1965-72]).

39 About Kleist see e.g. the paragraph in 'Ludwig Tiecks Dramaturgische Blätter,' *Goethes Werke*, Weimarer Ausgabe, XL, 178-79, and the conversation with J.D. Falk, probably in 1810 (No. 1373, ed. Biedermann; No. 3338, ed. Herwig). About Victor Hugo: 'Notre-Dame de Paris gelesen ... Es ist das abscheulichste Buch, das je geschrieben worden!' in a conversation with Soret, Weimar, June 27, 1831, as rendered by Eckermann (No. 2976a ed. Biedermann, Nachträge), see also Soret's version (No. 2976, ed. Biedermann; No. 6861, ed. Herwig); also the private journal of Maria Pawlowna, June 16, 1831 (No. 6854, ed. Herwig) and June 23, 1831 (No. 6859, ed. Herwig). About Mérimée's *La Guzla*, in a MS entry: 'der gräßliche Vampirismus' (Weimarer Ausgabe, XLII/1, 281-82). About Romanticism e.g.: Conversation with Riemer, Karlsbad, Aug. 28, 1808 (No. 1091, ed. Biedermann; No. 2731, ed. Herwig); 'Das Klassische nenne ich das Gesunde! und das Romantische das Kranke,' in a conversation with Eckermann, Weimar, April 2, 1829 (No. 2672, ed. Biedermann); 'pathologisch, oder auch romantisch,' in a conversation with Eckermann, Weimar, April 5, 1829 (No. 2674, ed. Biedermann); 'Classisch ist das

Gesunde, romantisch das Kranke,' in *Maximen und Reflexionen über Literatur und Kunst (Aus dem Nachlaß)*, Weimarer Ausgabe, XLII/2, 246, and similar statements on p. 247.

40 Mario Praz, *La carne, la morte e il diavolo nella letteratura romantica*, 3rd enl. ed., Contributi alle storia della 'civiltà europea,' (Firenza: Sansoni, 1948).

41 'La Morte amoureuse,' *La Chronique de Paris*, 23 and 26 June, 1836; later in Gautier, *Nouvelles* (Paris: Charpentier, 1845). 'Les Taches jaunes,' in Gautier, *Poésies complètes*, éd. René Jasinski, nouvelle éd. (Paris: A.G. Nizet, 1970), II, 242-44, No. 70.

42 August Graf von Platen, 'Tristan,' in *Sämtliche Werke*, historisch-kritische Ausgabe, eds. Max Koch and Erich Petzet (Leipzig: M. Hesse, s.d. [1910]), II, 94-95. ('Whosoever has seen beauty with his eyes,/ Is already a prey of death,/ He will not be able to serve life,/ And still he will tremble before death,/ Whosoever has seen beauty with his eyes.' My translation)

43 For the less immediately sexual death through love see Michel Fougères, *La Liebestod dans le roman français, anglais et allemand au dix-huitième siècle* (Ottawa: Naaman, 1974).

44 For Gusdorf see n. 1. Roland Mortier, *Clartés et Ombres du Siècle des Lumières* (Genève: Droz, 1969).

45 Roland Mortier, 'Lumières, préromantisme, romantisme: un problème de délimination,' in *Actes du VIIe Congrès de l'Association Internationale de Littérature Comparée: Montréal-Ottawa 1973*, eds. Milan V. Dimić and Eva Kushner, Bibliothèque de la *Revue Canadienne de Littérature Comparée*, 2 (Stuttgart: Kunst und Wissen, 1979), I, 105-107.

46 For this interpretation of Gothicism see Devendra P. Varma, *The Gothic Flame: Being a History of the Gothic Novel in England: Its Origins, Efflorescence, Disintegration, and Residuary Influence* (New York: Russell and Russell, 1966; 1st ed. 1957).

47 Walter Muschg, 'Goethes Glaube an das Dämonische,' *Deutsche Vierteljahrsschrift für Literaturwissenschaft und Geistesgeschichte*, 32 (1958), 321-43; Benno von Wiese, 'Das Dämonische in Goethes Weltbild und Dichtung,' in B. v. W., *Der Mensch in der Dichtung: Studien zur deutschen und europäischen Literatur* (Düsseldorf: Bagel, 1958), pp. 72-91.

2. 'Allusion' in the Eighteenth Century: The Disinherited Critic

It is a truth almost universally acknowledged in the twentieth century that literary allusion is one of the most characteristic and most expressive techniques of English literature in the eighteenth century, and we seem to have taken it for granted that a more or less explicit theory lay behind eighteenth-century practice. On examination, though, it turns out that eighteenth-century writers actually said very little about the function of allusion and that, faced with this disinheritance, we have had to attribute to them a theory based almost entirely on inferences from their practice. In this paper I should like to examine the little that the eighteenth century did say about allusion, suggest some possible reasons for what seems to be a great disproportion between theory and practice, and decide whether, as a result, we should revise our understanding of allusive practice during the period.

My subject is 'literary allusion,' that is, the technique whereby a writer composes his text so that it echoes and brings to the reader's awareness an earlier text which significantly affects the meaning of the later text. Among the many varieties of terminology that have been proposed for the two texts, I prefer to call the later one the alluding text and the earlier one the target text. One of the advantages of this latter term is that it implies activity and intention on the part of the author and thus helps to

0824-3298/84/0300-0023 $01.25/©C.S.E.C.S./S.C.E.D.S.

distinguish allusion from passive submission to external influence. The author aims the reader's attention at the target text.

'Literary allusion' can be a misleading term, though, because it may suggest an opposition to musical or cinematic allusion, for example, whereas these are actually closely analogous to literary allusion. Just as one literary text can allude to another, so musical compositions can allude to other musical compositions, films to other films. A given form can also allude to a neighboring form: conversation to literary works, a film to a painting. A more general definition of this kind of allusion might therefore read somewhat as follows: a technique whereby any system of signs ('text' in a broad sense) is made to recall a previously existing system of signs in such a way as to significantly affect its own meaning. The fundamental opposition is between this kind of allusion and the whole range of topical and historical allusions that can be grouped together, for want of a livelier term, as 'ordinary allusions' – that is, indirect or passing references to familiar objects, events, or persons in the real world – modern politicians, ancient generals, beautiful or otherwise notorious people, battles, strikes, scandals, etc. In this kind of allusion, sign systems refer to the real world, more or less directly; in 'literary' allusion there is also a reference to another sign system.

A brief example from Dryden's *Mac Flecknoe* can clarify this distinction. The second couplet of the poem reads

> This *Fleckno* found, who, like *Augustus*, young
> Was call'd to Empire, and had govern'd long:

A bit later, Fleckno declares

> Even I, a dunce of more renown than they,
> Was sent before but to prepare thy way;
> And coursly clad in *Norwich* Drugget came
> To teach the Nations in thy greater name.

In the first passage, Augustus is named directly, and Dryden expects his readers to know the facts about him that are relevant for the poem: his restoration of peace after civil war, his building programme that found Rome brick and left it marble, his patronage of outstanding poets like Horace and Virgil, and (just possibly) the despotic aspects of his rule.[1] The second passage reminds us first of similar phrases in certain biblical texts (especially Matthew 3:1-4 and 11:7-10); these phrases describe John the Baptist and thus bring him into the poem. This second method offers more scope to the reader's activity – Augustus is given to him, but he has to go looking for John the Baptist, even if not very far. These two kinds of allusion are clearly very different, but the use of the same term for both of them has caused considerable theoretical confusion, and it is only quite recently, especially in the work of Ziva Ben-Porat and Carmela Perri, of 1976 and 1978 respectively, that this basic distinction has begun to be explored.[2] Since my primary concern here is literary allusion, I will not consider, for example, the way Opposition writers alluded to Sir Robert Walpole and his alleged wrongdoings, whether by nicknames like 'brazen-face' and 'Screen-Master General' or by recounting episodes from ancient history that were very similar to episodes in his career and then blandly denying the possibility of any contemporary application.[3]

A certain degree of similarity between the alluding text and the target text is obviously necessary for a successful literary allusion, but a certain degree of difference is equally necessary — a difference of genre, subject, tone, or the like. Otherwise the later text would not have enough independent meaning for the earlier text to act upon, but would simply repeat in its own context the meaning of the earlier text. Creating and interpreting allusions requires both wit and judgment in Locke's sense; one needs to be able to see both similarities and differences.[4] From this situation arises the frequent comparison of the two texts of an allusion to the two terms of a metaphor, where both similarity and difference are also essential.[5] (It is worth reminding ourselves here that, however similar the two texts may be, there is no allusion unless the meaning of the later text is affected.)

The interplay between similarity and difference is obvious enough in mock-heroic allusions, but it operates in other kinds as well. Consider the second line of Pope's couplet:

> Laugh where we must, be candid where we can;
> But vindicate the ways of God to Man.[6]

The echo of Milton's statement of purpose in *Paradise Lost*, 'And justify the ways of God to men,' is perfectly clear, not only in the phrasing but in the position of each line as the conclusion of the opening paragraph of a long poem. If Pope's line occurred in a blank verse poem about the fall of men and angels, it would have little allusive force because the author would be trying, and probably failing, to do about the same thing as Milton. If it occurred in a twentieth-century prose treatise on the problem of evil, it would have little allusive force because the contexts would be too different. Here, the differences balance the similarities. Milton's line does not rhyme with the preceding line, whereas Pope's does. Milton's line begins with 'And' and follows a statement of another equally solemn purpose: 'That ... I may assert Eternal Providence,' whereas Pope's begins with 'But' and follows a statement of a much less solemn purpose. 'Vindicate' is close in meaning to 'justify,' but not identical. Thus Pope's line both expresses his own purpose directly and implies, indirectly, that this purpose has something but not everything in common with Milton's, in both form and content.

These distinctions are not very rigorous, and are not at all exhaustive, but they will serve as a general description of the idea that has implicitly informed much of our study of eighteenth-century literature during recent decades. Among many other contributions we may note Earl Miner's book on Dryden, Reuben Brower's on Pope, and Christopher Ricks' essay on both poets.[7] In view of my own bias toward theory, I should perhaps acknowledge that lack of an adequately explicit theory of literary allusion has not prevented these critics, and many others, from discussing it with great insight. They clearly know what it is.

As I have said, there is nearly universal agreement about the importance of literary allusion in our century. Particular interpretations may arouse two chief types of question: How much of his target text can a writer justifiably expect to allude to? What kinds and quantities of potential target texts can eighteenth-century writers and readers be expected to know? The late Earl Wasserman caused a number of raised eyebrows, and very few critics seem willing to extend the contexts of allusion as far as Wolfgang Rudat does.[8] More recently, some critics have been treating the much broader phenomenon of intertextuality in such a way that allusion as a distinctive technique virtually disappears,[9] but the most sustained critique of the role of allusion in the period comes from Irvin Ehrenpreis.[10] This critique is only one part of a discussion of the place of implicit meaning in Augustan literature and therefore makes no claim to be exhaustive, but as usual when he attacks conventional wisdom about the eighteenth century Ehrenpreis teaches us something even if we finally decide to maintain the convention.

Ehrenpreis finds two main weaknesses in most allusion critics: first, if there is more than one possible source for an allusion, they arbitrarily pick the most prestigious – Milton rather than Blackmore, Virgil rather than Boileau, Horace rather than a minor poet from the Greek Anthology; second, they seriously overestimate the erudition of the average eighteenth-century reader. Both arguments seem to me to call attention to important aspects of eighteenth-century allusion but to view them from the wrong perspective. In his first argument, Ehrenpreis has perhaps been victimized by his own erudition, and he fails to distinguish between allusion and passing imitation. If we can point to three or four possible sources for a passage, he asks, why should one be singled out as the focus of an 'allusion'? He seems to imply that the multiplicity of possible sources makes them all equally accidental, that if a text echoes six different earlier texts it cannot really allude to any of them. The rebuttal seems fairly obvious – in most cases, only the prestigious ancient is different enough from the modern to allow room for the allusion to operate. Also, since one of the most important functions of eighteenth-century allusion is to enable the author to make implicit moral judgments, prestige is an important aspect of the target text. On a more practical level, the best-known author is clearly the best target for an allusion – more readers know Horace than the minor Greeks. More fundamentally, allusion is not merely a matter of resemblance to an earlier text but of effect on the meaning of a later text; some resemblances are allusions and some aren't, just as some sound patterns reinforce meaning and some don't. When faced with a number of possible sources, it is not arbitrary but essential to center on the one or two that are important for the meaning of the later text. To be sure, Ehrenpreis does catch some critics taking their own wish for an allusion for the poet's deed, or using doubtful allusions to support dubious readings of a poem, but it doesn't seem to me that this argument lessens in principle the importance of allusion.

In the second argument, which is perhaps not quite consistent with the first, Ehrenpreis turns from the multiplicity of possible allusions to the eighteenth-century reader's limited ability to recognize any of them. The main support for his contention that Sir John couldn't read very well either is the footnotes that several writers appended to their poems.[11] If, for example, at Canto IV, line 1 of *The Rape of the Lock*, Pope felt obliged to footnote his allusion to Book IV, line 1 of the *Aeneid*, 'by far the best-known part of the best-known of all [classical] poems,' what are we to think of the level of literacy of his readers (p. 13)? At first glance this doubt seems more reasonable than Wasserman's confidence (pp. 27-28) that such notes are meant to stimulate the reader to search out unannotated echoes. When Ehrenpreis goes on to say that 'Dryden's notes to *Annus Mirabilis* have equally cruel implications,' though, he is on less firm ground. Here is how Dryden introduces these notes: 'In some places, where either the fancy, or the words, were his [Virgil's], or any others, I have noted it in the Margin, that I might not seem a Plagiary.'[12] Far from presuming to tell his readers anything they don't know, Dryden is covering himself by proclaiming his own awareness of those imitations that he believes his readers are likely to spot.[13] Like other writers we shall soon examine, Dryden is not concerned at all with allusion in our sense, but only with imitation. Another very heavily annotated work of this period is *The Pilgrim's Progress,* whose margins are sometimes solidly black with citations to the Bible. Bunyan's motive seems to have been to proclaim his scriptural orthodoxy and authenticity, to defend his use of fiction by showing that it rested on biblical fact; certainly no one would conclude from the presence of these notes that his readers didn't know their Bible! Pope's announcement of his notes to the *Dunciad* is probably more a part of his satire of his readers than a reliable indication of the real state of their literacy: 'The *Imitations* of the Ancients are added, to gratify those who either never read, or may have forgotten them; together with some of the Parodies, and Allusions to the most excellent of the Moderns.'[14] Pope's note to *The Rape of the Lock* IV.1 does seem embarrassing; yet we should recall that he tried to accommodate his female readers, very few of whom knew Latin and most of whom had probably not read Dryden's translation. In any case, the import of footnotes is much more complicated than Ehrenpreis grants, and I at least am unwilling to allow them to diminish the importance of allusion for eighteenth-century literature.

As we turn now to examine what eighteenth-century writers thought about allusion, we immediately face serious terminological problems. 'Allusion' was an extremely vague term in the eighteenth century. Besides its present meaning, it could also signify figure, pun, comparison, trope, metaphor, reference in a very broad sense, or merely imitation.[15] Unless the context provides a clue, there is seldom any way of telling whether the word carries anything like its modern sense or not. In Pope's statement quoted above, for example, there is no clear distinction between imitations, parodies, and allusions. A few more examples will deepen our awareness of the range and ambiguity of our term:

> The Great Art of a Writer shews it self in the Choice of pleasing Allusions, which are generally to be taken from the *great* or *beautiful* Works of Art or Nature; for though whatever is New or Uncommon is apt to delight the Imagination, the chief Design of an Allusion being to illustrate and explain the Passages of an Author, it should be always borrowed from what is more known and common, than the Passages which are to be explained.
>
> Allegories, when well chosen, are like so many Tracks of Light in a Discourse, that make every thing about them clear and beautiful. A noble Metaphor, when it is placed to an Advantage, casts a kind of glory round it, and darts a Lustre

through a whole Sentence: These different Kinds of Allusion are but so many different Manners of Similitude ...

Thus Addison in *Spectator* 421, part of the series on the pleasures of the imagination.[16] In the original printing of No. 205, Addison signaled the transition from a description of a vicious character as a scarecrow to a warning against 'the Shoals and Quicksands of Life' by writing 'To change the Metaphor.' For the reprint, 'Metaphor' was changed to 'Allusion,' with little discernible change in meaning (2.302). With no less confusion, Defoe, an author seldom included in discussions of allusion, calls *Robinson Crusoe* an 'allusive allegorick History.'[17] Richard Hurd writes of 'the folly of taking *resemblances* for *imitations*, in this province of *allusion*.'[18] Bolingbroke associates allusion with other outmoded kinds of discourse, as contrasted to the demands of the present for reason and fact: 'Men now require something more real than figure, more precise than allusion, and more particular than metaphysical abstractions.'[19] Johnson's well-known discussion of metaphysical imagery maintains the uncertainty: 'The most heterogeneous ideas are yoked by violence together; nature and art are ransacked for illustrations, comparisons, and allusions.'[20] Johnson also testifies, in a kind of pre-incarnation of Ehrenpreis, that the eighteenth century, like the twentieth, had its allusion-hunters: critics with telescopes

see with great clearness whatever is too remote to be discovered by the rest of mankind, but are totally blind to all that lies immediately before them. They discover in every passage some secret meaning, some remote allusion, some artful allegory, or some occult imitation which no other reader ever suspected.[21]

Since we don't know what kind of allusion Johnson has in mind here, his evidence cannot contribute to our present investigation.

Comments on particular poems may be a little clearer than these general statements. In a note to *Dunciad* 3.53, Pope says 'This has a resemblance to that passage in *Milton*,' from which he quotes three lines before continuing 'there is a general allusion in what follows to that whole passage' (TE, 5, 155). This seems promising, but we cannot be sure exactly what force the term had for Pope, and he offers no hint of what he sees as the significance of his allusion. Nor, again, does Richard Hurd: 'Learned writers of quick parts abound in these delicate allusions. It makes a principal part of modern elegancy to glance in this oblique manner at well-known passages in the classics' (*Works*, 2, 308). Even more mysterious is James Beattie. In Book I, stanza 44 of *The Ministrel* he refers to a 'long-winded tale' featuring 'hags, that suckle an infernal brood, / And ply in caves th'unutterable trade.' An asterisk directs us to the bottom of the page, where we read 'Allusion to Shakespear. *Macbeth*. How now, ye secret, black, and midnight hags, What is't you do? *Witches*. A deed WITHOUT A NAME. Macbeth. Act 4. Scene 1.'[22] Nowhere else in the poem does Beattie provide a signpost to an allusion, and in spite of great effort I've been unable to discover why he does so here.

Unless I've missed something essential, then, it seems clear that assembling examples of 'allusion' will not help us greatly to discover what the eighteenth century thought about the subject. We will explore some other avenues in a moment, but none of them will lead to anything like a comprehensive theory. It is puzzling, this contrast between the pervasiveness of the practice of allusion (unless we moderns have indeed greatly overestimated its importance) and the paucity of explicit discussion of it. Eighteenth-century critics certainly had at hand everything they needed to work out a theory. Apart from Pope's practice, there was Dryden's highly concentrated use of allusion in *Mac Flecknoe* and *Absalom and Achitophel*, not to mention many other less popular works. Before him, there was Milton. There was classical sanction in Virgil's use of Homer. There were well codified methods of reading the Bible typologically, rather weakened from their patristic and medieval origins, but still influential in some quarters.[23] Associationist psychology was ready with a convenient conceptual framework for explaining the operation of allusion. Yet David Hartley devotes a long chapter of his *Observations on Man* to 'Associations of Words' without coming close to allusion. Hartley was not particularly interested in literature, of course, but Alexander Gerard was in his *Essay on Genius* (London, 1774). He devotes several chapters to the various kinds and effects of the association of ideas, and he states that a single word can remind a reader of a large context, but he is interested only in memory, not in allusion:

> When memory presents one event, this leads us to think of the other events which were connected with it, and to conceive them in their proper order. The remembrance of many periods of a discourse or of a great number of verses, is often revived by the single word with which they begin. (p. 252)

Hugh Blair, Adam Smith, Lord Kames, Lord Monboddo are equally silent as far as I have been able to discover.

Another way into our subject that looks promising is via 'parody.' We have seen the term in Pope's preface, and will see below how Warton uses it in his *Essay on Pope*. Johnson's definition, in his dictionary, is useful: 'A kind of writing, in which the words of an author or his thoughts are taken, and by a slight change adapted to some new purpose.' This definition accounts for the difference between the two texts that is essential for allusion, and it avoids limiting the meaning of parody to ridicule of the text parodied, in striking contrast, for example, to the OED. Such a definition could provide a good starting point for a theory of allusion, but Johnson provides only one illustrative quotation: Pope's statement about the *Dunciad* notes in which, as we have seen, parody, allusion, and imitation are not clearly distinguished. Elsewhere, though, Johnson clearly shows his awareness of one way allusion can operate. In the well-known *Rambler* 168 he condemns the unsuitably low diction of a passage in *Macbeth:* 'dun' is a word used by stable hands, knives belong to a menial occupation, and, in general, 'words become low by the occasions to which they are applied, or the general character of them who use them; and the disgust which they produce, arises from the revival of those images with which they are commonly united.' He goes on to treat literary allusion, though without naming it, as a special case of the influence of past associations on the effect of words:

> Thus if, in the most serious discourse, a phrase happens to occur which has before been successfully employed in some ludicrous narrative, the most grave and serious auditor finds it difficult to refrain from laughter, when those whose imagination is not prepossessed by the same accidental association of ideas, are utterly unable to guess the reason of his merriment. (*Works,* 5, 126-27; quoted in the folio version, which is a bit more explicit.)

What makes this a real discussion of literary allusion is the focus on the meaning of the alluding text. Allusion from a serious text to a ludicrous one is less common in the eighteenth century than is allusion in the other direction, and when it does occur it is usually unintentional, but the operation is the same. Johnson's analysis also testifies to the importance of the association of ideas as a basis for allusion. Being primarily concerned with diction, though, he didn't carry his discussion of allusion any farther.

The Jonathan Richardsons, father and son, help a little and tantalize a great deal in their *Explanatory Notes and Remarks on Milton's Paradise Lost* (1734). Commenting on 4.778, where the cherubim leave heaven through an 'Ivory Port,' they remind us of Homer's two gates of sleep, of horn and ivory, through which respectively issue true and false dreams, and to which Virgil is thought to have alluded in *Aeneid* VI in order to imply that Aeneas's visit to the underworld is a poetical fiction. They argue that *if* Milton intended to allude to these predecessors, his purpose, like Virgil's, was to underline the fictiveness of his descriptions. They conclude by admitting that 'this is only Offer'd as a Conjecture; but most Certain it is He could not Forget this Ivory Gate of *Homer*, whether he Alluded to it or No' (pp. 178-79). This is the only place where the Richardsons treat an allusion as such, though, and the status of the gates of horn and ivory as famous symbols almost independent of their original texts perhaps lessens the value of their observation for our purposes. Richardson Sr. tantalizes us in his preface:

My Son has a very Copious Collection of [Milton's imitations], and as they often Assisted Us in Our Understanding Our author's true Meaning, they would (if inserted) have been Our Vouchers in Those Cases. To have added All These Fine Passages would have been Improper in Our Present Undertaking ... . (pp. cli-clii)

We cannot be sure how they would have related their copious collection to Milton's meaning, but we can wish that they had at least given us a few samples. We might expect help from Gilbert Wakefield's *Observations on Pope* (1796) because of his keen ear for echoes, but he doesn't point out the allusive significance of the sources he discovers. His preface seems to disclaim any such intention: 'My sole aim and desire has been to illustrate Pope, as an elegant English classic, by opening the sources of his imitation, by noticing his beauties of sentiment and expression, and occasionally his improprieties in both' (p. xvi).

Joseph Warton's *Essay on the Genius and Writings of Pope* adds a little material to our discussion.[24] A propos of *The Rape of the Lock*, he observes

The parodies are some of the most exquisite parts of this poem. That which follows from the 'Dum juga montis aper,' of Virgil [i.e., 3.165ff.], contains some of the most artful strokes of satire, and the most poignant ridicule imaginable. ... The introduction of frequent parodies on serious and solemn passages of Homer and Virgil, give much life and spirit to heroi-comic poetry. ... The mind of the reader is engaged by novelty, when it so unexpectedly finds a thought or object it had been accustomed to survey in another form, suddenly arrayed in a ridiculous garb. (1.243-45)

These expressions are satisfactory as an initial description of the general effects of Pope's allusions, but Warton doesn't try to show in detail how those effects are produced, and his attention seems focussed at least as much on the target text as on the alluding text. He discusses the *Dunciad* in a similar vein: 'The frequent parodies introduced on Homer, Virgil, Milton, and other great poets, than which nothing has a stronger effect in heroi-comic poems, are made with a singular pleasantry, happiness, and judgement' (2.378-79). When he turns to Pope's satirical use of the Bible, though, his entire attention is devoted to the effect of the parody on the target text, and he can only speak of 'profaneness' (2.380).

He comes closest to describing how allusion actually works in his comment on *An Essay on Man* I.16, which we examined above: 'Pope seems to have hinted, by this allusion to the Paradise Lost, that he intended his poem for a defence of providence, as well as Milton' (2.63). Even here, though, he concentrates on similarity to the neglect of difference. His succeeding comments show a clear awareness that Pope's purposes are different from Milton's, but he does not explicitly interpret the differences between the alluding text and the target text as signals of the difference of purpose.

Imitation is an essential aspect of allusion, and there is no shortage of eighteenth-century discussions of imitation. They turn out to be useful for us, however, less for what they have to say about allusion than for the way they steadily avoid the subject. Throughout this part of my essay I am greatly indebted to Roger Lonsdale's article, 'Gray and "Allusion": The Poet as Debtor.'[25] He has covered imitation very thoroughly, and any suspicions that I suffer from the anxiety of scholarly influence are justified. Throughout classical antiquity, the Middle Ages, and the Renaissance, imitation was regarded as a normal and necessary part of literary composition.[26] Only piracy – claiming another writer's work as one's own – and the most servile kind of close copying were generally condemned. The bee that gathers nectar from many sources – best known to us from its starring role in Swift's *Battle of the Books* – has a long and honourable history as an emblem of creative imitation.[27]

Imitation retained its place of honour in eighteenth-century theory. In his Preface of 1717 Pope declares that 'all that is left us is to recommend our productions by the imitation of the Ancients' (TE, 1, 7). A more remarkable discussion of imitation, one of the few relevant texts not mentioned by Lonsdale, occurs in *Guardian* 12:

> But over and above a just Painting of Nature, a learned Reader will find a new Beauty superadded in a happy Imitation of some famous Ancient, as it revives in his Mind the Pleasure he took in his first reading such an Author. Such Copyings as these give that kind of double Delight which we perceive when we look upon the Children of a beautiful Couple; where the Eye is not more charm'd with the Symmetry of the Parts, than the Mind by observing the Resemblance transmitted from Parents to their Offspring, and the mingled Features of the Father and the Mother.[28]

Here the earlier text, as an object of imitation, has a status virtually equivalent to that of Nature as Addison sees it in 'The Pleasures of the Imagination':

> This Secondary Pleasure of the Imagination proceeds from that Action of the Mind, which compares the Ideas arising from the Original Objects, with the Ideas we receive from the Statue, Picture, Description, or Sound that represents them. (*Spectator* 416; ed. Bond, 3, 559-60)

In contrast to most imitations, recognition of the ancient text is necessary here in order to achieve the double significance or 'new beauty' which is based on the relationship between the two texts. We also have a striking use of the metaphor of transmission and inheritance that Miner and Ricks have shown to be so fundamental in the poetics of Dryden and Pope (see note 7). It would seem to be only a step to a discussion of the impact of the earlier text on the meaning of the later one, but our author doesn't take that step. He does go on to state that allusions to the Bible or to ancient poets can add to the dignity of modern sermons or poems, but here, as in the passage quoted, he concentrates on similarity, on the way one text reproduces another.

Other Renaissance metaphors for imitation besides filiality and the bee recur in the eighteenth century. The earlier text is often figured as a mine or quarry from which the poet digs out materials, and one of a poet's merits is his ability to find material in unlikely places. Pope provides a satirical version of this image in the *Peri Bathous: 'Virgil* is said to have read *Ennius*, out of his Dunghil to draw Gold, so may our Author read *Shakespear*, *Milton*, and *Dryden*, for the contrary End, to bury their Gold in his own Dunghil.'[29] The success of an imitation often depends on the relative poetic merits of the borrower and the lender. Here the complementary metaphors are theft and military conquest, with negative and positive connotations respectively. Dryden justifies Ben Jonson's numerous borrowings in this way, granting him a kind of *droit du seigneur* over all the poetry produced in his domain:

> But *Ben* made nobly his, what he did mould,
> What was another's Lead, becomes his Gold;
> Like an unrighteous Conqueror he raigns,
> Yet rules that well, which he unjustly gains.[30]

One implication of all these metaphors is identical: once the excavated, or quarried, or conquered material is in place in a new text, its origin ceases to have any relevance for its meaning except perhaps as a testimony to the author's apian industry.

In spite of this theoretical sanction, however, as Lonsdale has shown so well, the practice of imitation produced great uneasiness in eighteenth-century poets. We have seen Dryden's defensive remarks in his preface to *Annus Mirabilis*. The youthful Pope inquired of his friend Walsh 'how far the liberty of *Borrowing* may extend.'[31] He was concerned about his pastorals, which are extremely imitative but not very allusive. He goes on to compare the borrowing writer to a branch grafted onto a tree of another stock so as to 'yield variety,' and then declares that 'a mutual commerce makes Poetry flourish; but then Poets like Merchants, shou'd repay with something of their own what they take from others; not like Pyrates, make prize of all they meet.' Wolfgang Rudat has made much of this passage as an emblem of Pope's allusive technique, but it is hard to see where the mutuality lies (see note 8). How does one repay Virgil for the lines one takes from him? By enshrining him in one's own poem as an act of homage? Perhaps, but the mercantile metaphor seems to me to intensify the air of doubt and uncertainty that the opening question reveals; again, there is no hint that the ancient text affects the meaning of the modern one. Later in his career Pope seems to have gone so far as to suppress a projected compilation of his imitations and parodies by William Bowyer.[32]

As the century progressed, this anxiety of imitation became more and more acute as more and more curious scholars investigated the borrowing habits of poets both ancient and modern. Critics worked hard to distinguish between real borrowings, coincidental resemblances, similarities determined by the nature of the subject, and other varieties of imitation and near imitation. Could vernacular poets be imitated as freely as the classics? Could contemporaries be imitated as freely as Elizabethans?

Two pervasive features of these discussions are important for our purposes: their forensic tone and the constant recurrence of competition and emulation as hallmarks of imitation. Pope himself provides an example of the latter in his definition of true wit as 'What oft was *Thought*, but ne'er so well *Exprest*' (*Essay on Criticism*, 298). It is prominent in the two dissertations that Richard Hurd published in 1751 as appendices to his edition of Horace: 'On Poetical Imitation' and 'On the Marks of Imitation.' Writers proclaim imitation, he says, 'when they have a mind to shew their dexterity in contending with a great original.'[33] Warton is more explicit in his *Essay on Pope*, whose pastorals, closely imitated from Virgil and Theocritus, are judged inferior to their models (1, 1-11). *Messiah*, though, 'incontestably surpasses the Pollio of Virgil' (11). At the beginning of his discussion of the Imitations of Horace, he mentions that Pope modified Horace's style by following that of Juvenal (2, 274), but thereafter nearly all his remarks are comparisons of the quality of corresponding passages in the two poets. In all such exercises he takes great care to be impartial:

As I before produced a passage of Milton which I thought superior to a similar one of Pope, I shall, in order to preserve impartiality, produce another from Milton, in which I think him inferior to the last quoted passage, except perhaps in the third line. (1, 26)

Behind all these judgments lies the assumption that the imitator is trying to do the same thing as his model and that the only question is who does it better. Once the judgment is made, the losing text can almost be discarded. Edward Young distinguishes imitation from emulation, but he also takes it for granted that in both cases the later poet is doing the same thing as the earlier; the difference is that the imitator is content to do less well than his model, whereas the emulator tries to do better.[34]

Forensic language sets the tone of the introductory question of Hurd's first dissertation, and throughout he conducts himself like a prosecuting attorney: 'Whether, in the case of confessed Imitations, any certain and necessary conclusion holds to the disadvantage of the natural GENIUS of the imitator?' (p. 110). 'Coincidences of a certain *kind*, and in a certain *degree*, cannot fail to convict a writer of *imitation*' (p. 214). 'Every degree of likeness is exposed to the *suspicion* of imitation' (p. 252). 'An *identity of phrase and diction*, is a much surer note of plagiarism ... There is no defending coincidences of this kind' (p. 208; Hurd's italics throughout). Yet by the end of the dissertation, we realize that Hurd is really acting as defence attorney, not as prosecutor: he seldom accuses writers of plagiarism, but rather defends them against anticipated accusations from his readers. He concludes this part of his argument:

It follows, from these considerations, that we cannot justify ourselves in thinking so hardly, as we commonly do, of the class of *imitators;* which is, now, by the concurrence of various circumstances, become the necessary character of almost all poets. Nor let it be any concern to the *true* poet, that it is so. For *imitations,* when real and confessed, may still have their merit; nay, I presume to add, sometimes a *greater* merit, than the very originals on which they are formed'. (p. 229)

Yet by constantly assuring us that imitation is normal, natural, and indeed inevitable, he betrays in himself the same anxiety that he finds in his contemporaries:

though many causes concur to produce a thorough degeneracy of taste in any country; yet the principal, ever, is THIS ANXIOUS DREAD OF IMITATION IN POLITE AND CULTIVATED WRITERS. (p. 240)

Warton is equally cautious, protesting, after a ten-page list of Pope's presumed borrowings from other writers,

I should be sensibly touched at the injurious imputation of so ungenerous, and indeed impotent a design, as that of attempting to diminish, or sully the reputation of so valuable a writer as Pope, by the most distant hint, or accusation of his being a plagiary. (1, 98)

He goes on to defend Boileau against similar imputations. So with William Clark in his projected annotations of Pope: far from making accusations, or even treating his discoveries as texts deliberately alluded to, he writes to Bowyer

I hope you do not suppose that Mr. Pope had half those passages in his thoughts that bear a resemblance with them. Poets, like painters, in this respect, are furnished with a collection of figures, images, drawings, descriptions, & c. which are always ready in their heads, without thinking of the original.[35]

Right in the middle of all these questions is Thomas Gray. Immensely learned, severely inhibited, obsessively imitative, he constitutes a crucial test-case for eighteenth-century allusiveness. Contemporary readers found scores of borrowings in his very small corpus, nineteenth-century editors added scores more, and Roger Londsdale's edition, another source of great indebtedness for me, lists literally hundreds in its massive footnotes.[36] Gray was both praised and censured by his contemporaries for the way he borrowed from his predecessors, and in his collected *Poems* of 1768, in Lonsdale's words, he

felt obliged to acknowledge for the first time a number of his 'imitations,' all of them notably acceptable according to traditional theory, some even ostentatious in their obscurity: they include Homer, Virgil, Lucretius, Dante, Petrarch, Spenser, Shakespeare, Milton and even Dryden but, with one exception, no later poet.[37]

Like Dryden's notes to *Annus Mirabilis*, Gray's are designed more to forestall carping critics than to instruct ignorant ones, though the latter motive is certainly present in Gray as well. This is borne out by a comment in an earlier letter about accusations of plagiarism: 'I could shew them a hundred more instances, wch they never will discover themselves.'[38] Gray sounds almost like the bank robber who accepts a jail term cheerfully, knowing that his securely hidden loot will be waiting for him when his sentence expires. As Lonsdale tells us, though, most of his additions to the body of Gray's borrowings come from the main line of English poetry since Spenser (p. 31). Ancients, moderns, Englishmen, Europeans, the famous, the obscure – Gray imitated everyone, eclectically and enthusiastically.

Not the least striking aspect of all these imitations is how few of them are allusive. The keenest allusion hunters, furnished by Lonsdale with an apparatus which makes the California Dryden and the Twickenham Pope look under-annotated, have found very little game. Harold Bloom mentions in passing that Gray is second only to Milton for 'allusive density,' but he must be using the term in a sense different from mine, and in any case he gives no examples and doesn't develop his pronouncement.[39] Someone has tried to find allusive significance in the way 'the crested pride / Of the first Edward' (*The Bard,* 9-10) echoes 'the crested adder's pride' from a song in Dryden's *Indian Queen,* III, i, but I am not convinced. Importing the meaning of 'adder' into Gray's text, even if a legitimate method, doesn't add much to the bard's denunciation of Edward, and the rest of Dryden's song is too commonplace to provide any enrichment. Thomas Gilmore tries to apply Earl Wasserman's methods to the Eton College ode, but the results, compared with what one finds in Pope, strike me as meagre.[40] Perhaps the reason is that allusion is too intellectual a method to communicate very well the delicate emotional tones evoked by this poem. The 'Ode on the Death of a Favorite Cat,' a much more Augustan performance with its sly humour and potential mock-heroics, might seem to offer more scope for allusion. In his essay, Lonsdale examines the candidates, from Helen of Troy to Virgil's Camilla, that have been proposed as allusive avatars of Selima the cat, and I must agree with him as he adopts Ehrenpreis's skeptical stance and concludes that none of them is very central to Gray's poem (pp. 48-54). There is no lack of parallels, echoes, and resemblances, but their effect on the meaning of Gray's poem is negligible. One of Gray's few real allusions occurs in his brief but mordant satire, 'On Lord Holland's Seat near Margate,' where line 16, 'mimic desolation covers all,' aptly echoes the close of the *Dunciad,* 'Universal Darkness buries All.'[41] The effect of the allusion is to associate Gray's little poem with Pope's much larger one, thereby implying that Lord Holland's estate is not an isolated aberration but part of a more general cultural collapse. Otherwise, it seems to me that the only allusive effect of Gray's many borrowings is a very generalized one, a kind of acknowledgment of his poetic predecessors and an implicit declaration that he is carrying on their tradition and is grateful for the materials they have supplied him with.

With one major exception: an emphatic, and very problematic, allusion that also involves the *Dunciad*. It is certainly unintentional, and it seriously damages his own text. In the third stanza of 'The Progress of Poesy,' Gray describes the infant Shakespeare ('thy' in the second line refers to England):

Far from the sun and summer-gale,
In thy green lap was Nature's darling laid,
What time, where lucid Avon strayed,
To him the mighty Mother did unveil
Her awful face:

These few lines contain loud echoes of three separate passages of the *Dunciad*, all of which refer to Dulness or to Colley Cibber. Their juxtaposition to Gray's lines shatters the mood of high sublimity that he was trying to maintain.[42] He has inadvertently made himself into a Dunce after the fact:

The Mighty Mother, and her Son who brings
The Smithfield Muses to the ear of Kings,
I sing. (1.1-3)

A veil of fogs dilates her awful face. (1.262)

But in her Temple's last recess inclos'd,
On Dulness' lap th'Anointed head repos'd.
Him close she curtains round with Vapours blue,
And soft besprinkles with Cimmerian dew. (3.1-4)

Nor is this discovery the work of a mischievous modern allusion-monger. It was noted in the eighteenth century by a writer in *The Gentleman's Magazine* who 'regretted that the reader was so "forcibly reminded" of Pope,' and by Gilbert Wakefield, who remarked in his 1786 edition of Gray's poems that 'wicked memory brings into the mind the *Queen* of the *Dunces*, and destroys all the pleasure of the description by an unlucky contrast.'[43] The situation here is similar to the one described in general terms by Johnson: the memory of a ludicrous text destroys the effect of a solemn one. There is clearly nothing ludicrous about Gray's diction as such. The relation between allusion and intention has occupied many critics, and most would deny the possibility of unintentional allusion, but in this case it is difficult to deny the destructive force of Pope's lines. Their effect, once they are pointed out, is perhaps all the stronger for being unexpected, since the reader of 'The Progress of Poesy' is not normally sensitized to the *Dunciad* the way the reader of the *Dunciad* is to Virgil and Milton. It is curious that an unintentional allusion by a habitually unallusive poet should have provoked, in Wakefield and in the anonymous gentleman, the most explicit eighteenth-century statement of the effects of allusion that I know about.

How could Gray have blundered like this? One can only speculate, but the likeliest answer seems to me to be that he simply didn't think allusively. That is, once he had appropriated an image or a phrase the original context ceased to be important. He saw it only in the context of his own work; he treated it possessively; the act of appropriation made it his own. The results were seldom so catastrophic as those we have just seen, because Gray borrowed mostly from sources whose contexts did not clash violently with his own, and from such a wide variety of sources that potential clashes cancelled each other. Thus the final curious aspect of this powerful but unintentional allusion is the evidence it provides of Gray's habitual unallusiveness, for no poet who usually expected his own text to be actively compared to an earlier text could slip like this.

How then can we account for the great disparity between our notion of the practice of eighteenth-century writers and their almost complete silence about the nature and function of allusion? Given the case of Gray, we might argue that the use of allusion decreases markedly after the mid-century. It is certainly less prominent then, but far from absent: Cowper, for example, has a complex and very important allusive relationship to Milton, and Blake's 'London' alludes forcefully to Ezekiel.[44] Another possibility is that allusion was so habitual a mode of writing and reading that critics felt little need to make its operations explicit, leaving it to act as an unexamined convention of discourse. This explanation would be more convincing if there were fewer eighteenth-century explanations of the obvious aspects of other topics. Warburton, for example, never shrinks from explaining features of Pope's verse that seem quite straightforward, yet he never tries to show exactly how his allusions operate.

I think a better explanation can be found in the way eighteenth-century critics responded to imitation and its discontents. Even when justified, or indeed inevitable, as in writing on timeless and universal subjects like love, duty, and sunsets, imitation was seen as a necessary evil, the result of a fall from an earlier state of poetic grace and originality. Any similarity between two texts is potentially incriminating; the only real guarantee of innocence is difference. The assumption that imitators are doing the same thing as their models and cannot avoid competing with them intensifies the anxiety. Similarity and difference become antithetical. A mind-set like this makes it very difficult to accept allusive writing as a deliberate imitation of an earlier text for a significantly different purpose. Recognition of the similarities blinded readers to the differences, and vice-versa. Questions of property come in also. When successful imitation is seen as appropriation or assimilation, when a text asserts its own identity only by destroying the identity of an earlier text, it is hard to understand a technique in which the identity of a later text depends on maintaining the separate identity of an earlier text while at the same time asserting that the two texts are similar, sometimes almost identical. Eighteenth-century readers were thus prevented from understanding explicitly what I think writers like Dryden and Pope understood intuitively: the positive role that allusion can play in a fallen poetic world by enabling a poet to put to work, for his own purposes, the virtues and strengths of predecessors whom he cannot escape anyway.

To what extent, then, should these considerations modify our approach to eighteenth-century allusion? Do they justify Ehrenpreis's scepticism? Have the methods of reading that we have developed for Eliot and Joyce distorted our approach to Dryden and Pope? If we choose a historicist approach and seek eighteenth-century justification for allusive reading, I think there is enough evidence, scrappy and unsatisfactory as it is, to show that our interest in allusion is not entirely anachronistic. Their much greater preoccupation with imitation need not deter us from concentrating on allusion, unless we take the extreme position that our explicit critical categories should correspond exactly to those of the period we are studying. Understanding why they did not develop a more explicit rationale for allusion can help us relate our interests more clearly to theirs. The absence of allusion in Gray does not call into question its presence in Pope and Dryden; if anything, their allusiveness stands out all the more clearly by contrast.

Nevertheless, some cautions are in order. As I hope I have shown, not all writers are equally allusive. Not every imitation is an allusion. We have to guard against the temptation to inflate that afflicts all scholars: a 'source' is good only for a paragraph in *Notes and Queries*, but an 'allusion' is worth at least a substantial note and perhaps a whole article. Finally, we need to develop more precise and rigorous definitions of allusion, its functions, its capabilities, its limitations, and its relation to other poetic techniques. Here, instead of lamenting that our eighteenth-century forebears have disinherited us, we should perhaps rather rejoice that they have not pre-empted us.

WILLIAM KINSLEY
Université de Montréal

Notes

1 See Howard D. Weinbrot, *Augustus Caesar in 'Augustan' England* (Princeton, NJ: Princeton University Press, 1978) for an exhaustive account of the negative side of Augustus' image. Even a seemingly straightforward historical allusion like this one may have ambivalent meanings.

2 Ben-Porat, 'The Poetics of Literary Allusion,' *PTL*, 1 (1976), 105-28, and her Berkeley dissertation, 'The Poetics of Allusion' (1973); Perri, 'On Alluding,' *Poetics*, 7 (1978), 289-307.

3 For a thorough treatment of this aspect of Pope, see Maynard Mack, *The Garden and the City: Retirement and Politics in the Later Poetry of Pope, 1731-1743* (Toronto: University of Toronto Press, 1969), esp. chapters 4-5.

4 *An Essay Concerning Human Understanding*, 2.11.2.

5 This aspect of allusion is illuminated by Maynard Mack in ' "Wit and Poetry and Pope": Some Observations on his Imagery,' in *Pope and his Contemporaries*, ed. James L. Clifford and Louis A. Landa (Oxford: Clarendon Press, 1949), pp. 20-40.

6 *An Essay on Man*, 1.16. Pope's poetry is quoted from the Twickenham Edition of his *Poems*, gen. ed. John Butt (London: Methuen; New Haven: Yale, 1939-69), hereafter abbreviated as TE.

7 Miner, *Dryden's Poetry* (Bloomington: Indiana University Press, 1967); Brower, *Alexander Pope: The Poetry of Allusion* (Oxford: Clarendon Press, 1959); Ricks, 'Allusion: The Poet as Heir,' in *Studies in the Eighteenth Century III*, ed. R.F. Brissenden and J.C. Eade (Toronto: University of Toronto Press, 1976), pp. 209-40.

8 Among Rudat's many articles, see especially 'Allusive Technique in Pope's Early *Rota Virgilii* Poetry,' *Antike und Abenland*, 22 (1976), 70-99; 'Another Look at the Limits of Allusion,' *Durham University Journal*, 71 (1978), 27-34. For reactions to these and other articles, see the critiques in *The Scriblerian*. I agree that Rudat usually goes too far, but it is difficult to fix the limits of allusion unless someone with his combination of ingenuity and erudition risks going beyond them.

Wasserman, 'The Limits of Allusion in *The Rape of the Lock*,' *JEGP*, 65 (1966), 425-44. For a critique, see Ehrenpreis (note 10). See also reviews of Wasserman's *Pope's 'Epistle to Bathurst': A Critical Reading with an Edition of the Manuscripts* (Baltimore: Johns Hopkins University Press, 1960) by Robert M. Schmitz, *PQ*, 40 (1961), 418-20, and by Maynard Mack, *MLN*, 76 (1961), 872.

9 For one example, see Gérard Genette, *Palimpsestes: La littérature au second degré* (Paris: Seuil, 1982). Peter Hughes' stimulating attempt to go beyond conventional notions of allusion seems to me premature because we don't yet understand even

the conventional notions well enough: 'Allusion and Expression in Eighteenth-Century Literature,' in *The Author in his Work: Essays on a Problem in Criticism*, ed. Louis L. Martz and Aubrey Williams (New Haven, CT: Yale University Press, 1978), pp. 297-317.

10 *Literary Meaning and Augustan Values* (Charlottesville: University of Virginia Press, 1974), pp. 1-25. Ehrenpreis's more recent book, *Acts of Implication: Suggestion and Covert Meaning in the Works of Dryden, Swift, Pope, and Austen* (Berkeley: University of California Press, 1980) unfortunately discusses allusion only in passing, e.g. on pp. 88-89.

11 'Either' is a signal to look elsewhere, and I hope that some readers, at least those of my generation, caught the echo of Rudolf Flesch's once-famous book, *Why Johnny Can't Read* (New York: Harper & Row, 1955). For what intention is worth, the allusion is meant to imply that 'illiteracy' is a perennial problem.

12 *Poems*, ed. James Kinsley (Oxford: Clarendon Press, 1958), 1, 48.

13 This was also one of Eliot's many motives for adding notes to *The Waste Land*: 'I had at first intended only to put down all the references for my quotations, with a view to spiking the guns of critics of my earlier poems who had accused me of plagiarism' ('The Frontiers of Criticism,' *On Poetry and Poets* [London: Faber and Faber, 1957], p. 109).

14 TE, 5, 3rd ed. (1963), 9. Cf. the more satirically convoluted dance of the same terms on pp. 52-53.

15 For a copious compilation of eighteenth-century dictionary definitions of 'allusion,' see Howard D. Weinbrot, 'The "Allusion to Horace": Rochester's Imitative Mode,' *SP*, 69 (1972), 350-52.

16 *The Spectator*, ed. Donald F. Bond (Oxford: Clarendon Press, 1965), 3, 578.

17 He uses this phrase in his *Serious Reflections* on Robinson. See Robert W. Ayers, '*Robinson Crusoe:* "Allusive Allegorick History",' *PMLA*, 82 (1967), 399-407.

18 'A Dissertation on Poetical Imitation,' *Works* (1811), 2, 196-97. Hurd's italics.

19 'Essay the First, Concerning the Nature, Extent, and Reality of Human Knowledge,' *Works* (Philadelphia, 1841), 3, 131.

20 'Life of Cowley,' *Rasselas, Poems, and Selected Prose*, ed. Bertrand H. Bronson, 3rd ed. (San Francisco: Rinehart, 1971), p. 358.

21 *Rambler* 176, *Works* (New Haven, CT: Yale University Press, 1958 –), 5, 167.

22 *The Minstrel*, ed. Pierre Morère, Publications de l'Université des Langues et Lettres de Grenoble (1981), p. 94.

23 See Paul J. Korshin, *Typologies in England 1650-1820* (Princeton, NJ: Princeton University Press, 1982), and Thomas R. Preston, 'From Typology to Literature: Hermeneutics and Historical Narrative in Eighteenth-Century England,' *The Eighteenth Century: Theory and Interpretation*, 23 (1982), 181-96.

24 4th ed. (1782; rpt. Farnborough, Hants.: Greg, 1969). Cited parenthetically in the text by volume and page.

25 In *Studies in the Eighteenth Century IV*, ed. R.F. Brissenden and J.C. Eade (Canberra: Australian National University Press, 1979), pp. 31-55. Lonsdale's subtitle alludes to Ricks's essay (see note 7), Oxford don saluting Cambridge don from

far-off Australia, and in my subtitle a professeur de l'Université de Montréal salutes Oxbridge from Alberta.

26 See Harold Ogden White's thorough survey, *Plagiarism and Imitation during the English Renaissance* (Cambridge: Harvard University Press, 1935) and the more analytical study of G.W. Pigman III, 'Versions of Imitation in the Renaissance,' *Renaissance Quarterly*, 33 (1980), 1-32.

27 Pigman, pp. 4-7.

28 Pope, *Prose Works 1711-1720*, ed. Norman Ault (Oxford: Blackwell, 1936), pp. 90-91. One of Ault's reasons for attributing this essay to Pope is the resemblance of some of its images to those of the Preface of 1717, but Harold Williams calls the evidence for the attribution 'meagre' (*RES*, 13 [1937], 490).

Hurd finds a similar sort of pleasure in imitation, but he too fails to see any effect on the meaning of the later text: 'The *pleasure* we find in this exercise of the imagination ['to set two objects together, and contemplate their mutual aspects and appearances'] is the main source of that perpetual usage of *indirect and allusive imagery* in the writings of the poets' (*Works*, 2, 195).

29 *The Art of Sinking in Poetry*, ed. Edna Steeves (1952; rpt. New York: Russel and Russel, 1968, chap. 9, p. 39.

30 Prologue to *Albumazar*, 11-14; *Poems*, ed. Kinsley, 1, 141.

31 Letter of 2 July 1706; *Correspondence*, ed. George Sherburn (Oxford: Clarendon Press, 1956), 1, 19-20.

32 Lonsdale, p. 38; John Nichols, *Literary Anecdotes of the Eighteenth Century* (1812-15; rpt. New York: AMS and Kraus, 1966), 4, 429-37.

33 *Works* (1811), 2, 274. For a different view of 'allusion as agon,' see Robert Folkenflik, ' "Homo Alludens" in the Eighteenth Century,' *Criticism*, 24 (1982), 223-25. For earlier theories of emulation, see Pigman, 'Versions of Imitation.'

34 'Conjectures on Original Composition,' *Complete Works*, ed. James Nichols (1854; rpt. Hildesheim: Georg Olms, 1968), 2, 568-69. For a brilliant anatomizing of Young's essay, see Joel C. Weinsheimer, 'Conjectures on Unoriginal Composition,' *The Eighteenth Century: Theory and Interpretation*, 22 (1981), 58-73.

35 Nichols, *Literary Anecdotes*, 4, 429.

36 *The Poems of Thomas Gray, William Collins, Oliver Goldsmith* (London: Longmans, 1969).

37 'Gray and "Allusion",' p. 45.

38 Letter to Edward Bedingfield, 27 Aug. 1756; cited by Lonsdale, p. 45.

39 *A Map of Misreading* (New York: Oxford University Press, 1975), p. 125.

40 'Allusion and Melancholy in Gray's *Ode on a Distant Prospect of Eton College*,' *Papers on Language & Literature*, 15 (1979), 52-59.

41 Gray's line is even closer to the ending of the 1729 *Dunciad*, 'universal Darkness covers all.'

42 On these lines see also John E. Sitter, 'Mother, Memory, Muse and Poetry after Pope,' *ELH*, 44 (1977), 312-15.

43 *Gentleman's Magazine*, 51 (1781), 569; Lonsdale, *Poems*, p. 172.

44 Dustin Griffin, 'Cowper, Milton, and the Recovery of Paradise,' *Essays in Criticism*, 31 (1981), 15-26; Harold Bloom, *Poetry and Repression* (New Haven, CT: Yale University Press, 1976), pp. 34-45. See also James K. Chandler, 'Romantic Allusiveness,' *Critical Inquiry*, 8 (1982), 461-87.

3. Gulliver Shifts for Himself

Sartorial tropes are common devices in literary and graphic satires. Ann Messenger's recent essay on clothing imagery in the Henriad traces the satire on tailors, tailoring and fine forms of words back at least as far as the political, philosophical and religious controversies of the English Renaissance.[1] Ronald Paulson has associated the foppery of the three brothers in *A Tale of A Tub* with such ludicrous figures as Sir Fopling Flutter, with the travesties of the Restoration stage and graphic caricature.[2] The limits of the sartorial trope are the shame of nakedness and the absurdity of pomp, which correspond to the bluntness of plain-speaking and the incomprehensibility of euphemism or euphuism. There are symbolical coats, such as Gilray's impoverished John Bull often wears, with trailing tags of referential identity. Such are the allegorical coats of Martin, Jack, and Peter in Swift's most famous sartorial foray – allegorical coats, at least according to the learned gloss by Wotton and the tub-thumping Hack. But there are also analogical coats of satire, fabricated by analogical tailors. Instead of symbolizing the products, these mimic the processes of expression and euphemism, thought and delusion.[3] The Swiftian version of this analogical trope of clothing-and-words is rather less obvious and famous than the symbolical clothes-philosophy and the three brothers' self-tailoring in *A Tale of A Tub*. The missing analogical tailor is Lemuel Gulliver. What he tailors is a strait-jacket of ideas.

The subject of the analogical trope of clothing-and-words – misinterpretation of symbols, confusion of symbol with referent, of vehicle with tenor – reminds us of the particular early occasions of the analogical satire: abuses of the eucharist, of dogma, and of the crown. Studies of

MAN AND NATURE / L'HOMME ET LA NATURE III/1984
0824-3298/84/0300-0047 $01.25/©C.S.E.C.S./S.C.E.D.S.

the Swiftian tailor by Paulson, Philip Harth, Miriam Starkman, and Frances Louis have concentrated on these occasions, have shown that much of the tailoring even in *A Tale of A Tub* is really analogical, and have defined the realm of unknowing which extends beyond the immediate objects in the satire. It is a realm where accidents usurp essence as the object of knowing, where phenomena pass for noumena, where knowing becomes a mere choice of nice delusions. The denizens of this philosophical mirror-world regard clothing as the true body, words as the nominal reality and exclusive reality. Their truth is a philosophy of clothes, of superficialities – that is, a clothing to hide their philosophical nakedness. The tailor, like the rhetor, lays claim to true philosophy. But his admirable work merely covers and falsifies the body, as the rhetor covers his subject with embellishments which become an end in his art.

In the analogical satire of clothing-and-words, the donning, doffing, and tailoring of clothes follows the ways of expression, interpretation, and thought, and the clothes themselves are not particularly symbolic. Harth, Paulson, and, lately, Louis and Frederick Smith, have treated *A Tale of A Tub* as a critique of impure reason.[4] Harth compares the tailor-worship and the philosophy of a universe of clothing to the Aeolists' worship of 'wind' as universal principle, and he sees both absurd religions as parodies, hence analogies, of reductive monism in general. Harth comments on the scope of the analogical satire: 'In ridiculing a form of atheism as enthusiastic madness through the reductive system of the tailor-worshippers, [Swift] was simply adopting and turning to his own use a well-established convention of Anglican rationalist polemics against atheists.'[5] According to Harth, the immediate target of Swift's sartorial analogy is the opening argument of Hobbes' *Leviathan*, an argument also analogical in its method; the tailor-worshippers' terminology clearly resembles Hobbes'.[6] The two branches of the tailor-worship to which Swift refers are two kinds of materialism.[7] One kind denies all but the material, sensible reality: if man and the universe are but a suit of clothes, there is no body beneath it (i.e. 'spirit' is really phenomenal). Phenomenon is the only noumen, or else noumen is null. The second kind is traducianism. Its doctrine, in Swift's version, is that 'man was an animal compounded of two dresses, the natural and the celestial suit, which were the body and soul: that the soul was the outward, and the body the inward clothing; that the latter was *ex traduce;* but the former of daily creation and circumfusion.'[8] The materialists' dependance upon sensations, upon 'the films and images that fly off upon their sense from the superficies of things,'[9] leads them to 'reduce the whole to one of its infinitesimal parts – a part which has a real existence but is of a very limited character.'[10]

Paulson unravels the sartorial analogy even more meticulously. He argues that the object of Swift's sartorial satire in *A Tale of A Tub* is the Gnostic heresy in particular, with an extension to the methods of interpretive and dogmatic self-sufficiency practised in all heresies. Though Paulson grants that the business of the three brothers' coats is allegorical, as Wotton and later glossers allege,[11] the tendency of Paulson's interpretation of the whole sartorial business in the *Tale* is to show allegory yielding to analogy. The first mention of the brothers' inheritance and its perversion brings to the Hack's mind a complete philosophy of clothes, and it is this doctrine or method which the embellishment and reformation of coats illustrates. Passing from the allegorical notion of the coats as 'a representation of an external law, the "visible church" — that by virtue of which one is not oneself but part of a community,'[12] Paulson takes up the meaning of the act of embellishment.[13] He moves from the symbolism of coats to the analogy between tailoring and the operations of heretical philosophy. On the analogy of the Renaissance Christian view that the 'body is the vesture of the soul ... dress is not adornment but part and parcel of the self, an indication of what one *is*. Ideally, dress and self are mutually informing; but the clothes philosophy to which the brothers yield makes the Gnostic distinction ... between body and soul (reminiscent of the Ramist distinction between form and content). Neither body nor soul can be said to bear any responsibility for the other ... The tailor represents the separation ... between outer and inner, the tendency of the Gnostic mind to think that it can cut the exterior ... to the size of its own vision ... whether by altering the institutional aspects of Christianity or by stripping them away altogether and going naked ... or, in more general terms, that it can change the style with the season.'[14]
Louis extends this treatment of *A Tale of A Tub* as a critique of unreason to *Gulliver's Travels*,[15] but she does not develop the analogy between Gulliver's tailoring and his unknowing. It might seem that Swift's amusement with the sartorial trope ended with the virtuoso performance in *A Tale of A Tub*. But there is plenty of tailoring in *Gulliver's Travels*, and it is entirely analogical, without the allegorical distractions (or the burlesque of the allegorizing pedant) of the *Tale*.

In *Gulliver's Travels* there are at least three kinds of tailors. Our attention, however, is not fixed upon their products, as it is upon the three brothers' inherited, elaborated, and reformed coats. We attend, instead, to the means of craftsmanship. The Lilliputian tailors fashion a patchwork container for Gulliver; their looms will make no broader cloth. The tailors of Laputa take Gulliver's measurements indirectly, through plane geometry and trigonometry. In the land of the Houyhnhnms, there are no native tailors, and Gulliver must shift for himself.

Among the expensive provisions for Gulliver's comfort in Lilliput are the Emperor's orders that 'three hundred tailors should make me a suit of clothes after the fashion of the country.'[16] In what will become a recurrent matching of the sartorial and the linguistic, these orders are immediately followed in the regal protocol by orders for Gulliver's tutoring in Lilliputian. When Gulliver and his Majesty are able 'to converse together in some sort' (I, 2, 27) Gulliver's first words 'express my desire that he would please to give me my liberty.' The Emperor frames a polite yet dodgy reply, and begs Gulliver's indulgence 'if he gave orders to certain proper officers to search me; for probably I might carry about me several weapons, which needs must be dangerous things, if they answered the bulk of so prodigious a person' (I, 2, 27). Gulliver is indeed a prodigy for the Lilliputians, yet the Emperor's plan to search his clothing for dangerous weapons shows at once the Lilliputians' acknowledgment and misunderstanding of the prodigy. The body of Gulliver, ignored in the Emperor's polite proposal, is weapon enough, as Gulliver's exploits against the Blefuscan fleet will prove. We see that Gulliver, however, has already put on the cognitive 'fashion of the country,' before the sartorial, for, in order to satisfy his Majesty, he is 'ready to strip myself, and turn up my pockets before him.' Yet he can express this offer only 'part in words, and part in signs' (I, 2, 27). Here, as elsewhere in his account of the travels, Gulliver asserts yet contradicts his linguistic abilities: he recalls, improbably, exact translations of phrases which, at the time he writes about, he could barely utter. The episode closes with a long, finicky inventory of Gulliver's pockets.

The whole episode resolves into patterns of analogy, between the wearing of clothes and the wearing of a system of thought, and between both of these acquisitions and the acquisition of a language. The Lilliputians busy themselves about Gulliver's pockets, but they haven't a proper respect for his person. Moreover, because of the vast differences, between Gulliver and the Lilliputians, in scale of conception, the Lilliputians misconstrue most of what they find in Gulliver's pockets. In the pattern of internal analogies of *Gulliver's Travels*, such misconstruction of parts-and-whole, of details-and-forms, foreshadows Gulliver's dread of the naked female form, whether Brobdingnagian or European or Yahoo, a dread attached first to the stripped serving-women in the court of the Brobdingnagian Queen, a dread later extended to Gulliver's own stripped essence, among the Houyhnhnms.

Gulliver confidently proceeds to trade niceties in an alien tongue, but he doesn't realize that in putting on its forms of words he is also putting on its forms of thought. Thus he proudly shows himself in the fashion of Lilliput well before he describes the tailoring of his garb, and despite the fact that the fashion is too small to suit him. Even his offer to strip in order to prove his harmlessness is part of a complicated ironical pattern, like the pattern by which the misconstruction of 'skins' is foreshadowed or analogically reproduced. When the Lilliputians don't take up the offer, Gulliver preserves his privacy by refusing the inspectors entry into three of his pockets, one of them purposefully secret. The latter contains 'a pair of spectacles (which I sometimes use for the weakness of my eyes), a pocket perspective, and several other little conveniences; which, being of no consequence to the Emperor, I did not think my self bound in honour to discover' (I, 2, 30). Gulliver hides the artificial means of sharp vision, perhaps excusing this concealment differently when he remarks of the Lilliputians that 'their sight is much more acute than ours' (I, 2, 30). The spectacles and perspective may suggest near and far sight, as well as the pun on 'spectacle.' Gulliver thinks them of no use *to another*, but he also appears never to use them with proper skill. Later, Gulliver is thought to have stripped when he has not. The Brobdingnagian farmer, first beholding Gulliver, mistakes Gulliver's coat for his skin; in other words, he thinks that Gulliver is naked. In contrast, Gulliver takes pains to avoid stripping before his master, or any, Houyhnhnm. On the two occasions when his nakedness is exposed in Houyhnhnmland, the results, as we shall notice, are philosophically disastrous.

Unlike the puns upon the tailor's yard and goose in *A Tale of A Tub* which emphasize sexual prowess or rapacity,[17] the figures of the tailor and of tailoring in *Gulliver's Travels* always emphasize the manner of plying the needle, and especially the tailor's incompetence and obliviousness to failure. The other kind of copula-tion is at issue — the engendering of a garment: of words, of concealment. By the method of tailoring we understand the method of philosophy in each realm. The three hundred royally-appointed tailors in Lilliput, for example, may be able to make clothing for Gulliver superficially 'after the fashion of the country,' but neither the finished garments nor the method of manufacture could possibly follow that fashion. It is plain that clothing Gulliver is a major national project, no ordinary task of tailoring. Gulliver keeps careful tally of the manpower devoted to tending him, and the three hundred tailors set to the task may well remind us of the strenuous engineering project of capturing and transporting Gulliver in the first place. The results of both projects are absurd. The Lilliputians can no more encase than really bind Gulliver. Through neither project do they really apprehend him. The packthreads of his captors are better suited to sartorial than military uses, at least according to human scale. The rope of the political rope-jumpers appears to Gulliver merely another such packthread. The 'two hundred sempstresses' who sew his shirts and linen employ materials 'all of the strongest and coarsest kind they could get; which, however, they were forced to quilt together in several folds, for the thickest was some degrees finer than lawn' (I, 6, 51). Unlike the Laputan tailors, the Lilliputians do bother with Gulliver's measurements, but their ingenuity of plumb-line, ladders, and literal rule-of-thumb is another instance of audacious efforts and puny results: 'my clothes ... looked like the patchwork made by the ladies in England, only that mine were all of a colour' (I, 6, 51). Gulliver apparently rests content with these garments, just as he tests no further the fiction that he is a captive of the Lilliputians. The patchwork garments 'all of a colour' are, like the co-operation necessary to make them, a deceptive, ambiguous matter. They define not only the smaller scale of Lilliputian technology but also the Lilliputians' limited vision, their mastery only of parts, their pettiness. The Lilliputians co-operate to subdue and tend Gulliver, but their natural tendency is factional. They and the Blefuscans wear violently uncomplementary colours of the same religious fabric and differ, of course, about the correct height of heels. When they isolate Gulliver as a faction or cabal of one, a proposal for his assassination is that they poison his shirts and bedclothes.[18]

After the Brobdingnagians discover the difference between Gulliver's skin and his clothing, they try to suit him. The task requires microsurgical dexterity, because Gulliver is tiny even for a Brobdingnagian doll. But the emphasis in this case falls upon the strength of Brobdingnagian cloth, the tailor's success in managing his unaccustomed task, and Gulliver's awkwardness of adaptation: upon the playfulness, not the audacity, of the tailor's effort. Glumdalclitch, the amateur, makes a batch of doll clothing from the finest fabric, which feels to Gulliver 'coarser than sackcloth.'[19] The Queen's tailor fares slightly better, working with silks 'not much thicker than an English blanket, very cumbersome till I was accustomed to them. They were after the fashion of the kingdom, partly resembling the Persian, and partly the Chinese, and are a very grave decent habit' (II, 3, 84). The discomfort seems Gulliver's, not his hosts'. We suspect that the gravity and the decency of his habit (intended qualities of his rhetoric, too) amuse his hosts, as an animated doll's pretensions might amuse us. Gulliver is quite literally being toyed with.

In short, not the making but the wearing of the clothes is the problem, from the second voyage onward. In Brobdingnag the problem is clearly Gulliver's. Thus the sartorial objects of satire are fixed not in the alien but in the human realm, where they remain for the rest of the Voyages. Most of the occurrences of clothing imagery in the Voyage to Brobdingnag are occurrences of Gulliver's false pride or misconception of himself. Gulliver loses his suit of clothes when he tumbles into a cream bowl, pushed by a dwarf. On four other occasions Gulliver befouls his clothing; on the last of these, significantly, while showing off by trying the long-jump over a cow-dung.[20] Because the radical shift in perspective has put Gulliver on the same scale as frogs, moles, and flies, he cannot keep up outward signs of value: he cannot keep clean. Yet the association of clothing and self-respect takes another turn by the end of the second voyage. Among other evidences of his life as a toy, Gulliver displays proudly 'the breeches I had then on, which were made of a mouse's skin' (II, 8, 119). Considering that, to the Brobdingnagians, Gulliver and the mouse are of equal stature, this display gives a foretaste of Gulliver's uses of Yahoo-skins and of his categorical confusions.

The Laputans and Gulliver regard each other's appearances with shared astonishment. Gulliver describes Laputan 'outward garments' (III, 2, 127) – an odd yet suggestive redundancy. The garb is 'adorned with the figures of suns, moons, and stars, interwoven with those of fiddles, flutes, harps, trumpets, guitars, harpsichords, and many more instruments of music, unknown to us in Europe' (III, 2, 127). Laputan dishes, he discovers, are similarly carved or contrived (III, 2, 129). The Laputans wear and eat the forms of their conceptual fascinations. Not only are these forms symbolic of inward 'garb' or thought; the methods of fashioning both inward and outward garb are also analogous. We see the analogy in Laputan tailoring. The Laputan tailor measures Gulliver by survey and, it would seem, without touching him. After six days of creation, the tailor brings 'clothes very ill made, and quite out of shape, by happening to mistake a figure in the calculations' (III, 2, 130). This parody of the Creation and of the origination of cosmic evil escapes Gulliver's understanding. He pretends to take comfort from observing 'such accidents very frequent and little regarded' and he soons offers as further palliative the parallel instance of Laputan houses, 'very ill built, the walls bevil, without one right angle in any apartment' (III, 2, 131).

Gulliver explains that 'this defect ariseth from the contempt they bear for practical geometry, which they despise as vulgar and mechanic ... although they are dextrous enough upon a piece of paper in the management of the rule, the pencil, and the divider, yet in the common actions and behaviour of life I have not seen a more clumsy, awkward, and unhandy people, nor so slow and perplexed in their conceptions upon all other subjects, except those of mathematics and music' (III, 2, 131). The limitation is also linguistic: 'Imagination, fancy, and invention, they are wholly strangers to nor have any words in their language by which those ideas can be expressed ...' (III, 2, 131). Gulliver, however, doesn't explicitly assemble these conclusions by coupling unhandiness with thoughtlessness or philosophical error. His own habits of mind don't incline him toward such a comparison. Yet accidents are 'little regarded' in Laputa because Laputans have little regard. The Laputan tailors cannot make simple coverings *because* they proceed through calculations abstracted from the body. The attempt at figuring (without an underlying figure) is the point of the satire, not the mistake in figuring, which is an irrelevant excuse. Gulliver recognizes the faulty, ludicrous products but not the fallible method of production. There is no right fit of the artificial and the natural, partly, it seems, because the linguistic mediation or pattern is missing. The Laputans, we recall, are singularly oblivious to linguistic signs, hence their need for 'flappers' to signal the very fact of utterance. It should not surprise us that Gulliver witnesses the people of Lagado dressed 'generally in rags ... a people whose countenances and habit expressed so much misery and want' (III, 4, 142). Nor that the first scientific projector whom Gulliver meets in the Academy has 'clothes, shirt, and skin ... all of the same colour,' the colour of soot (III, 5, 145). His madness is visible; he is all of a piece, shameless in asserting his crazed thoughts and in panhandling from Gulliver.

In the land of the Houyhnhnms, neither the rational horses nor the Yahoos are tailors. The Yahoos lack shame and art, the motive and the means. The Houyhnhnms have no conception of clothing. Gulliver bewilders the Houyhnhnm philosophers not only because he seems a clothed Yahoo, a contradiction in terms, but because he is clothed at all, thus a term beyond Houyhnhnm knowledge. When he first beholds a Yahoo and starts to suspect, in horror, a likeness of species with his own, Gulliver notes with fear and relief that the Houyhnhnms do not perceive the likeness between his feet and the Yahoo's 'because of my shoes and stockings' (IV, 2, 186). In contrast, though in similarly delusive philosophical manner, Gulliver is busy ignoring significant differences, such as hairiness, color, skin texture, and length of nails. Gulliver's wearing of gloves is especially puzzling. The master Houyhnhnm signs for him to remove them and 'reduce' his hands 'to their former shape' (IV, 3, 187). This order to expose the reduced shape of the hand is followed immediately, in the now-recognizable sartorial-linguistic match, by an order 'to speak the few words I understood.' In the master's estimation of Gulliver's linguistic prowess there is also a reduction. The juxtaposition of commands is particularly significant in light of the later disclosure that the Houyhnhnms, who have no reason for clothing and therefore tailor none, have no reason for lies and therefore can utter none. For Gulliver's first night among the Houyhnhnms, the master 'ordered a place for me to lodge in; it was but six yards from the house, and separated from the stable of the yahoos. Here I got some straw, and covering myself with my own clothes, slept very sound' (IV, 2, 188-189). Gulliver is properly placed apart from both Houyhnhnms and Yahoos, safely slumbering, for the time being, under his own coverings.

But Gulliver's separateness from the Yahoos is ambiguous, as his philosophy of categories breaks down under the pressure of a radically different system – and, no doubt, under the disability of his own unhandiness as philosopher. The master Houyhnhnm 'was convinced ... that I must be a yahoo, but my teachableness, civility, and cleanliness astonished him; ... He was most perplexed about my clothes, reasoning sometimes with himself, whether they were a part of my body; for I never pulled them off till the family were asleep, and got them on before they waked in the morning' (IV, 3, 189). Other Houyhnhnms cannot believe Gulliver 'to be a right yahoo, because my body had a different covering from others of my kind' (IV, 3, 190). Gulliver aggravates the mystery of species, by keeping 'the secret of my dress, in order to distinguish myself as much as possible from that cursed race of yahoos' (IV, 3, 191). There is a crisis of philosophy, of self-knowledge, when Gulliver is accidentally exposed – while asleep, *without* his own coverings. A servant Houyhnhnm reports to the master 'that I was not the same thing when I slept as I appeared at other times; ... some part of me was white, some yellow, at least not so white, and some brown' (IV, 3, 191). Like his Lilliputian suit, Gulliver's body appears a patchwork – but a patchwork of several shades. When naked he does not seem whole. Of his deliberate self-concealment Gulliver remarks: 'I found it in vain to do so any longer' (IV, 3, 191). The context enforces a pun upon 'in vain' – vanity and futility. Practical Gulliver foresees the latter sense, ignores the former. Soon his clothes will wear out and he will have to tailor for himself, 'by some contrivance from the hides of yahoos or other brutes' (IV, 3, 191). The matching words 'hides' and 'brutes' protect Gulliver from dim suspicions of species kinship and complicity in genocide.

Gulliver decides to explain clothing to the master Houyhnhnm. Thus he accounts for the tailor's art: 'in the country from whence I came those of my kind covered their bodies with the hairs of certain animals prepared by art, as well for decency, as to avoid inclemencies of air both hot and cold' (IV, 3, 191). He offers to demonstrate the difference between clothing and flesh 'if I did not expose those parts that nature taught us to conceal' (IV, 3, 191). The master, perplexed by the whole idea of clothing, is perplexed especially by Gulliver's claims of decency and natural shame. Houyhnhnms follow the perfectly rational moral imperatives of nature; the nature that dictates shame to Gulliverian yahoos must appear mad. Upon close examination of clothing and flesh, the master Houyhnhnm concludes that Gulliver 'must be a perfect yahoo,' despite certain notable exceptions. Yet he allows Gulliver to put on his clothes again, 'for I was shuddering with the cold' (IV, 3, 192). This last sign that clothing may not be merely affectation doesn't influence the master's conclusions, which he repeats in the council held to decide Gulliver's fate in the proposed genocide. In both episodes, Gulliver's sartorial and linguistic traits are noted in tandem. After the master agrees to keep secret the fact of Gulliver's clothing, he next urges that Gulliver 'go on with my utmost diligence to learn their language, because he was more astonished at my capacity for speech and reason than at the figure of my body, whether it was covered or no' (IV, 3, 192). In council, the master mentions clothing and power of speech as first and second items on his list of Gulliver's traits. But just as the master's philosophy has no proper category for Gulliver, so it cannot accommodate the real analogy of clothing and language. The master Houyhnhnm treats speech and reason as a perfect correspondence; he cannot understand how speech may falsify reason, as clothing expands, decorates, conceals the body beneath, as clothing provides secret places. With thought, as with the body, it *does* matter 'whether it was covered or no.'

The Houyhnhnms' conceptual difficulties with lying and their arguments against the practice, including the argument of their own virtuous example, have often been taken as a simple satire upon human vice. But the sartorial trope suggests otherwise. The master Houyhnhnm's clothing-philosophy, like his philosophy of truth and lies, is adequate to the Houyhnhnm way of life, which is not, and cannot be, the human way. And not all of Gulliver's defence of clothing is unsound. His shivering is eloquent. The Houyhnhnms' philosophy of language – holding that language is pure communication because its motives are univalent – is itself a lie about the human reality – and perhaps not even a 'noble lie' such as men would do well to believe. If it is true for Houyhnhnms it is unattainable for men, who need their clothing in order not to freeze or broil to death. Gulliver's lies and disguises are odious, of course, but the Houyhnhnms' absolute truthfulness requires a Houyhnhnm's impervious hide, or a perfect comfort with one's native reality. Swift's very method in *Gulliver's Travels*, with its multiple ironies and elaborate fictions of authorship and editorship and narrative voice, requiring laborious and significant deconstruction, also gives the lie to Houyhnhnm linguistic philosophy. In a perfect example of that debunking irony at work, Swift has Gulliver report the closing of the master's treatise on the vacant art of lying thus: 'And these were all the notions he had concerning that faculty of lying, so perfectly well understood, and so universally practiced among human creatures' (IV, 4, 194). Whose smugness is this? Is it not the master Houyhnhnm pretending to be master, too, of the very subject of his ignorance. 'All the notions he had' are notions enough to dismiss the 'faculty.'

The simplicity of the Houyhnhnms' life captivates Gulliver. Over and over he invites us to admire the reduction of ethical complications and conceptual spider-webs to the simple, though static, network of the Houyhnhnms' thoughts and deeds. A Houyhnhnm can ply a needle for show, improbably, but doesn't need the craft; the Houyhnhnms 'have no letters' or alphabet for they have no literature to record. Like Laputans, they are devoid of imagination. Unlike Laputans (or Gulliver), they are also devoid of artifice. The general satirical theme of Chapter VI of the Fourth Voyage is the constrast between European luxuriousness and the natural, unimaginative, unlettered, untechnological existence of the Houyhnhnms. But the particular occasion of the contrast falls within the sartorial trope and therefore subtly alters the direction of the satire. An eleven-year-old Yahoo female sexually attacks Gulliver when she spies him naked, and she howls when he dons his clothing again. This incident is immediately followed by Gulliver's strongest admiration of the non-speculative, practical, ethical philosophy of the Houyhnhnms. At the moment of sharpest anguish and deepest philosophical confusion, Gulliver seeks a borrowed clarity, which he cannot have.

Gulliver says that at home he carries on his body 'the workmanship of an hundred tradesmen' (IV, 6, 204). The remark brings to mind Swift's frequent attacks, in *Gulliver's Travels* and in less heavily 'clothed' satires, against imported finery and wasted wealth.[21] But it also reminds us of the mutual dependencies within society — dependencies which the returned, over-compensating Gulliver can no longer feel or accept with clear conscience. The obverse of the vaunted Houyhnhnm simplicity is that it is easy to lose any member of the community.[22] When Gulliver must rely upon his own workmanship alone, when he must become his own tailor acording to his own patterns, he works with the skins of creatures whom he partly mistakes for his kin.

Finally, the sartorial trope figures in the most vexatious and important philosophical questions. As is usual with Swift, these questions are chiefly ethical, Is it right to treat creatures variously according to their kind? Are the categories or kinds real? What light reveals them and guides action aright? For ethical purposes, which are the accidental, which the essential, qualities of creatures? Gulliver's clothing somewhat confounds his hosts as they dimly acknowledge the force of such questions. The accidents of the body may similarly confuse our understanding of the spirit which wears the body as its 'outward garb.' The Houyhnhnms' social hierarchy, their scale of ethical value, corresponds perfectly to their various breeds and hues.[23] Racism is wisdom in Houyhnhnmland, where rank is really qualitative. Judgment by 'superficies of things' is wisdom. Gulliver must explain, ironically, that such a method is most unwise when one judges men, among whom the figure of the body belies the figure of the soul. Healthy appearance, among men, betokens vulgarity; weakness or disease betokens nobility of rank. The Yahoos suffer from disease; the Houyhnhnms, perfectly integrated, do not.

The master Houyhnhnm tenders his own clothes-philosophy. It touches upon the fate of the Yahoos and of Gulliver, and it reveals the intellectual 'fashion of the country' in a light not quite as flattering as Gulliver's wholesale admiration and emulation. The Houyhnhnm concedes that he once thought that clothing might be a wise provision for men, in order to prevent hatred among them caused by the sight of bodily deformities. If it preserved social harmony, clothing would be a noble lie. But the master rejects this concession because the source of evil lies deeper — in Yahoos, by his own observation, in men, by Gulliver's account of European politics and history. Hatred is not an aesthetic revulsion. The viciousness of the Yahoos, which results in wars and competition for unworthy objects, like human viciousness arises from 'gross defects in reason, and by consequence, in virtue' (IV, 7, 209). Gulliver would clothe these defects in a fine form of words. 'In order to favour' men, he has 'concealed many particulars, and often *said the thing which was not*' (IV, 7, 209). Here the master's philosophy fails him. For Gulliver's euphemisms for vice, though excessive and often pathetically futile, nevertheless work in exactly the way that the master would have accepted for clothing. The lies by which Gulliver seeks to ingratiate himself with his present readers not only justify mankind to an alien race; they are also the peaceable lies of community. Remove them altogether and man's nakedness is unbearable; men hate what they behold, even in themselves. Gulliver's 'present' state of antinomian confusion is fair proof of this hypothesis. We remember as well the most infamous victims of anatomy — that radical sort of untailoring — in *A Tale of A Tub:* the flayed woman and the dissected beau.[24]

Such stripping kills what it reveals, reverses what Carlyle admits as 'the benignant efficacies of concealment.'[25] When Gulliver accepts, admires, and patches into his own, the unhuman, inconsistent philosophy of the Houyhnhnms, he comes to hate men. He even falls into the absurdity of preferring the tender mercies of the savages who wound him without provocation to the generous charity of 'European yahoos' who would rescue him.

The Portuguese crewmen of the rescue ship wonder at Gulliver's garb, which, like his shoes, sails, and canoe, he has tailored from the skins of mere brutes: rabbits and yahoos have the same standing as raw materials for this tailor's art.[26] Of this tailoring Gulliver says, with chilling placidity: 'Necessity is the mother of invention'; of its aftermath, that he 'enjoyed perfect health of body and tranquility of mind' (IV, 10, 223). He has his wish – he has become a cool spouter of platitudes, a human parody of Houyhnhnm-hood. But for the tranquility Gulliver pays in confusion, shame, and loathing upon his return to human society. The signs of his condition – and of the shaky sanity of his borrowed philosophy – are largely sartorial. Gulliver refuses to undress aboard the rescue ship. He allows himself to be wrapped up in the Captain's cloak for concealment when they arrive in port, and he accepts a gift of the Captain's fresh shirts, but he refuses to let the Captain's tailor measure him. He says that he fears contamination by Yahoos. But we may suspect that he also fears what the measuring will reveal about his proper figure. It is clear to the rescuers that Gulliver has acquired outlandish garb, foolish ideas, and a barbarous language. Yet Gulliver takes pride in his tailoring. It is a mark of his independence from merely human standards and devices. This very pride, however, is an imperfect imitation of Houyhnhnm-hood, for the horses' language and philosophy hold no notion of pride. Finally, Gulliver admits that 'it be hard for a man late in life to remove old habits' (IV, 11, 238). The sartorial pun cuts several ways: at the truly 'old habits' of human pride and self-delusion; at the new habits (become old, i.e. trusted) which Gulliver has fashioned for himself; at his 'outward' and inward garb; at his determination to live as a Houyhnhnm, without protective illusions. Like the outraged Swiftian voice behind the projector of 'A Modest Proposal,' Gulliver has suited himself with a garment painful to wear and painful to remove.

MARK MADOFF
Royal Roads Military College

Notes

1 Professor Ann Messenger, of Simon Fraser University, has kindly permitted me to read her unpublished paper, ' "I'll murder all his wardrobe, piece by piece": Clothing Imagery in *I Henry IV*.'

2 Ronald Paulson, *Theme and Structure in Swift's Tale of A Tub* (New Haven, CT: Yale University Press, 1960), p. 27. See also Paulson's reference to the figure of the crazed tailor in *The Duchess of Malfi*, IV, 2 (p. 110 n) and to Carlyle's development of a dualism of body-and-clothes in *Sartor Resartus*, where, according to Paulson, 'a strong preference is shown for the naked body (the soul) over the clothing (the body).'

3 Paulson (p. 99) carefully unravels the obscure weaving of the pseudo-numinous in the 'Hebrew-like' gibberish of the Gnostic epigraph to *A Tale of A Tub,* with its arbitrary, impressive 'translation' attached: 'On the threshold of the *Tale,* it is a symbol of the sufficiency which goes so far as to create its own religion and language.' Paulson directs a student of the sartorial *analogy* to Swift's burlesque treatise *On the Mechanical Operation of the Spirit,* wherein a quilted cap prevents evaporation of perspiration except by mouth (words=excreta): '... communication can be either the result of a digestive process or of a mental process. The "Manufactory Operation" of the tailor and the "drawing and spinning" of the spider both suggest the mindless nature of the former' (p. 119).

4 Philip Harth, *Swift and Anglican Rationalism: The Religious Background of "A Tale of A Tub"* (Chicago: University of Chicago Press, 1961); Paulson, op. cit.; Frederick N. Smith, *Language and Reality in Swift's "A Tale of a Tub"* (Columbus: Ohio State University Press, 1979); Frances D. Louis, *Swift's Anatomy of Misunderstanding: A Study of Swift's Epistemological Imagination in "A Tale of A Tub" and "Gulliver's Travels"* (New York: Barnes & Noble, 1981); Miriam Starkman, *Swift's Satire on Learning in "A Tale of A Tub"* (New York: Octagon, 1968).

5 Harth, p. 80.

6 Harth, p. 85.

7 Harth, pp. 82-83.

8 Jonathan Swift, *A Tale of A Tub,* in *Gulliver's Travels and Other Writings,* ed. Louis A. Landa (Boston: Houghton Mifflin-Riverside, 1960), p. 283 (Section II).

9 Edward Stillingfleet, *Origines Sacrae* (7th ed.; Cambridge, 1702) quoted in Harth, p. 140.

10 Harth, p. 142.

11 Paulson, p. 27.

12 Paulson, p. 106.

13 'The Gnostics ... created a metaphor by simply giving a thing a name that belongs to something else. By this system any name and object can be associated at will' (Paulson, p. 142). Paulson draws examples of this procedure of 'secret analogy' from the Hack's metaphors (p. 143), and explains (p. 142) that one of the three intellectual methods required for reading the *Tale* sanely is a sense of 'common analogy' or of public, not esoteric, metaphorical significances.

14 Paulson, pp. 109-110.

15 Louis's valuable discussion of epistemological analogies in both satires centres upon the likenesses between physical size, percedptual apparatus, magnanimity, and ethical wisdom.

16 Jonathan Swift, *Gulliver's Travels,* in *Gulliver's Travels and Other Writings,* ed. Louis A. Landa (Boston: Houghton Mifflin-Riverside, 1960), p. 26 (I, 2, 26). All subsequent citations from *GT* will be given within parentheses in the text, in the format (Book, Chapter, page number in Landa's edition).

17 Paulson (p. 199) explains 'goose' as sexual favor or sexually available woman, 'yard' as sexual organ, 'seamen' obviously as its homophone, and 'needle' as an undermining description of the size of the phallus.

18 *GT*, I, 7, 56.

19 *GT*, II, 2, 77. This case of relativity in perception (what seems fine in texture to the Brobdingnagians seems coarse to Gulliver) does not deny the stability of the object of sense. But when the objects of sense are animate (e.g. persons), the epistemological crux is complicated by the intrusion of *blame:* '... a very ofensive smell came from their skins; which I do not mention or intend to the disadvantage of those excellent ladies, for whom I have all manner of respect; but I conceive that my sense was more acute in proportion to my littleness. ... Upon this point, I cannot forebear doing justice to the Queen my mistress, and Glumdalclitch my nurse, whose persons were as sweet as those of any lady in England' (II, 5, 94). The final comparison ironically leaves open the question of blame and of the stability of sensory objects.

20 The cream-bowl incident occurs at II, 3, 87. In the incident of the marrowbone (II, 3, 87-88) Gulliver soils his stockings and breeches. He ruins his suit through a fall into a mole-hill at II, 5, 94, and is befouled by frog slime at II, 5, 97. The cow-dung leap occurs at II, 5, 100.

21 A striking and famous example of the damning list occurs at IV, 6, 203: 'But in order to feed the luxury and intemperance of the males, and the vanity of the females, we sent away the greatest part of our necessary things to other countries, from whence in return we brought the materials of disease, folly, and vice, to spend among ourselves. Hence it follows of necessity that vast numbers of our people are compelled to seek their livelihood by begging, robbing, stealing, cheating, pimping, forswearing, flattering, suborning, forging, gaming, lying, fawning, hectoring, voting, scribbling, star-gazing, poisoning, whoring, canting, libelling, free-thinking, and the like occupations: every one of which terms, I was at much pains to make him understand.' The list is luxuriously illustrative, yet there are holes in its fabric: 'voting,' 'star-gazing,' and 'free-thinking' are odd entries. Having elaborated so many unfamiliar terms, for unfamiliar deeds, Gulliver both takes and suffers pains to interpret.

22 IV, 9, 221. 'Friendship and benevolence are the two principal virtues among the Houyhnhnms, and these not confined to particular objects, but universal to the whole race. ... They have no fondness for their colts or foals, but the care they take in educating them proceeds entirely from the dictates of reason' (IV, 9, 216). Gulliver registers this stoicism coolly. If he admires it, he does so quietly, through understatement.

23 The confusion of accident and essence is plainly a feature of the debate in the grand council of the Houyhnhnms to adopt a modest proposal about the Yahoo/Gulliver problem. The arguments have the manner of formal logical soundness, but the propositions about Gulliver's body, intellect and kind do not yield valid conclusions. Houyhnhnm racism (IV, 6, 207) matches Houyhnhnm eugenic marriage customs (IV, 8, 217).

24 Swift, *A Tale of A Tub,* ed. Landa, p. 333. The Hack's verb connects clothing and dissection: 'Yesterday I ordered the Carcass of a *Beau* to be stript in my Presence; when we were all amazed to find so many unsuspected Faults under one Suit of Cloaths: Then I laid open. ...' Dissection is a continuation of stripping.

25 Thomas Carlyle, *Sartor Resartus* (London: Dent, 1908), p. 174 (Book III, Chapter III, 'Symbols'). I am grateful to Professor Murray J. Evans of the University of

Winnipeg for directing me to an analysis of the ethics of truth-telling and concealment which resembles my own. In his letter to Eberhard Bethge of 5 December 1943, Dietrich Bonhoeffer turns to the sartorial-linguistic analogy as a means of ethical clarification: 'After all, "truthfulness" does not mean uncovering everything that exists. God himself made clothes for men; and that means that *in statu corruptionis* many things in human life ought to remain covered, and that evil, even though it cannot be eradicated, ought at least to be concealed. Exposure is cynical, and although the cynic prides himself on his exceptional honesty, or claims to want truth at all costs, he misses the crucial fact that since the fall there must be reticence and secrecy.' — *Letters and Papers from Prison*, ed. Eberhard Bethge (enlarged edition, London: SCM Press, 1967), p. 158. See also pp. 212-213, and the remark (p. 288): 'There are few people who know how to value reticence,' and Bonhoeffer, *Ethics*, ed. Eberhard Bethge (London: Fontana, 1964), Part Two, Chapter V, 'What is Meant by "Telling the Truth"?'

26 *GT*, IV, 10, 223 & 227.

4. Répétition et instabilité: la signification de *Zadig*

Dans l'introduction à son *Essai sur Candide*, Jean Sareil nous signale qu'aussi paradoxal que cela puisse paraître, deux siècles d'une abondante exégèse n'ont pas réussi à clarifier la signification d'un conte qui se veut naïf et simple. Les nombreuses interprétations de *Zadig* permettent de dire autant de ce conte. Il est donc avec beaucoup d'humilité que nous présentons ces quelques pensées sur *Zadig*.

La question centrale à tous les débats sur *Zadig* peut être réduite à un seul mot: 'Mais.' Faut-il prendre Voltaire au sérieux et croire que la leçon du conte est que nous sommes tous les marionnettes de la Providence et que nous devrions nous résigner à notre destin inaltérable? Ou faut-il plutôt croire que Voltaire se moque de l'idée exprimée par Pope et Leibnitz et selon laquelle chaque événement fait partie de l'ordre de l'univers, et selon laquelle tout mal entraîne un plus grand bien? Ou est-ce que le conte illustre le dilemme de Voltaire qui doit reconnaître l'existence ou la co-existence du mal et du bien sans pouvoir concevoir comment ils s'accordent?

Pour soutenir ces théories, certains critiques se sont servis de preuves d'ordre biographique et ont cité alternativement les prétendues réussites et les déceptions de l'auteur à l'époque où il rédigeait le conte. D'autres critiques se sont efforcés de montrer par le développement ou le non-développement du personnage principal, par la structure ou par l'absence de structure, par l'unité ou la dualité du complot, par l'ironie ou par son absence, que l'une ou l'autre des trois interprétations évoquées l'emporte.

0824-3298/84/0300-0063 $01.25/©C.S.E.C.S./S.C.E.D.S.

Il y a cependant un grand mal dont souffrent unanimement toutes ces interprétations: aucune n'est assez compréhensive pour éviter ou pour expliquer les contradictions apparentes que nous nous permettrons de relever brièvement.

Van den Heuvel traite *Zadig* en *Bildungsroman*.[1] Selon lui, Voltaire montre de plus en plus de sympathie pour ses personnages qui souffrent et évoluent. Zadig souffre, il est vrai, et il est également vrai qu'il est et roi et heureux à la fin. Mais le bonheur, comme cette mise en parallèle d'un substantif et d'un adjectif quasiment antithétiques, est instable. Nous constatons que chaque fois que Zadig se croit heureux, il est sur le point de devenir malheureux. Zadig, s'il a appris une seule chose, a appris à redouter les serpents lorsqu'il jouit d'un lit de roses. Plutôt qu'un Zadig heureux, nous voyons un Zadig prêt à retomber dans le malheur. Le lecteur est, pour ainsi dire, conditionné, après dix-neuf leçons, à ne plus s'en étonner. ... D'ailleurs, si Zadig doit son bonheur, comme il le dit, aux leçons qu'il a retenues: celle de l'ermite devenu ange et celle du voleur devenu riche, il est guidé par des idées contradictoires: on doit se soumettre au destin et on ne doit pas l'accepter. Si telle est donc la sagesse de Zadig, il est difficile de croire que l'élève est l'objet de la sympathie de l'auteur. N'oublions pas non plus que Zadig a énormément pitié de lui-même: il se croit le plus malheureux des hommes, au point de s'estimer, séparé qu'il est de son amour, cent fois plus malheureux que ce pêcheur qu'il rencontre et qui a perdu sa femme, sa maison, son commerce et qui est un homme tellement pitoyable que même les poissons (comme les hommes d'ailleurs) se moquent de lui. Chose bizarre, Zadig nous donne toujours la liste de ses malheurs, mais oublie de mentionner les bonnes choses ou n'en compose simplement pas la liste lorsqu'il est heureux. Comme Pangloss, il répète sa liste de malheurs et attribue mal les effets aux causes. Mais Pangloss est convaincu de sa philosophie et rien ne l'ébranle. En revanche, Zadig n'a que sa foi religieuse pour le soutenir, et ses malheurs et son manque de compréhension font qu'il murmure contre son destin apparent et contre la Providence. Mais revenons à la question de sympathie: cet homme qui dort si bien et si tard qu'il se fait voler, qui refuse de suivre les conseils de son ami, qui est si inefficace, si incapable d'agir et si prompt à se plaindre, cet homme est plutôt l'objet de notre pitié.

Il est également difficile de croire, avec Roland Barthes, à un Zadig idéalement bon du début à la fin.[2] Très tôt Voltaire nous dit de Zadig: 'Il était aussi sage qu'on peut l'être: car il cherchait à vivre avec des sages.'[3] La sagesse n'est pas une maladie contagieuse; on peut vivre longtemps avec des sages et rester parfaitement sot. D'autre part, qui sont et où sont ces prétendus sages? La compagnie de Zadig se compose de vaniteux, d'ingrats, d'infidèles, de jaloux et d'envieux: Sémire et Orcan, par exemple. Voltaire nous dit également que Zadig 'n'ignorait pas les principes physiques de la nature, tels qu'on les connaissait alors, et savait de la métaphysique ce qu'on en a su dans tous les âges, c'est-à-dire fort peu de chose' (p. 3). Zadig avait aussi 'de grandes richessess' et 'par conséquent, des amis' (p. 3). Il nous semble que Zadig manque d'amis, de métaphysique, de sagesse et même d'amour, passion qu'il ne découvrira qu'avec la jarretière de la femme de l'envieux et les rubans jaunes de la reine. (Les symboles de l'amour et de la jalousie sont interchangeables.) Si Zadig parle plusieurs langues et fait preuve d'une bonne interprétation de la justice, il est tout de mème totalement superficiel dans son raisonnement et complètement aveugle quant à ses rapports avec autrui. Zadig l'imparfait philosophe admet d'ailleurs qu'il ne comprend pas le discours des autres philosophes. Il ne se doute nullement des actions des jaloux et des envieux. Il murmure contre la même Providence qu'il nie. Enfin, il ne semble ni idéalement bon ni parfaitement sage.

Zadig n'est cependant pas totalement idiot comme l'ont suggéré d'autres critiques.[4] Il est toujours celui des hommes qui mérite le plus d'être éclairé, ce qui n'est pas très flatteur pour le lecteur.

Nancy Senior divise notre conte en deux: l'histoire du héros parfait, celui qu'épouse Astarté, et l'histoire du philosophe qui commence par questionner timidement et qui finit par tout questionner, même le discours de l'ange Jesrad.[5] Mais Senior admet que la solution n'est guère parfaite, car il y a des épisodes qui ne contribuent ni au développement de Zadig le sage, ni à celui de Zadig l'amoureux. Ces épisodes sont, conclut-elle, tout simplement amusants. Mais cet amoureux parfait ne tombe amoureux qu'au beau milieu du conte et ne se rappelle son amour qu'à la fin des chapitres, au moment où, après une autre aventure, il cherche l'Astarté de ses rêves sans pourtant la reconnaître lorsqu'il la retrouve. L'amoureux parfait n'est guère touché par Almona qui lui sauve la vie et lui offre son coeur. Il préfère voler la femme d'un autre, tout comme Orcan lui a pris Sémire. On pourrait d'ailleurs se demander s'il y a de l'amour ou seulement de la jalousie dans ce conte.

Zadig ne ressemble pas non plus à un philosophe parfait. Il se prosterne à l'envolée de l'ange et sa pire hérésie est de 'murmurer contre la Providence' qu'il finit par adorer (répété deux fois) et bénir (le dernier mot) (p. 59). Et si Zadig se permet de murmurer contre la Providence, c'est dans un état de choc ou de dépression, non pas à la suite d'un discours ou d'un soliloque philosophique, chose à laquelle il ne participe qu'une fois (lors du souper) et dont la conclusion (la Providence existe) ne nous révèle pas si cette Providence est bonne ou mauvaise.

Mais est-ce que le conte nous apprend que nous sommes libres ou que nous sommes victimes de notre destin? La réponse se trouve, presque tous les critiques le disent, dans le chapitre sur l'ermite.[6] L'ange donne à Zadig la triple illustration du fait qu'il n'y a point de hasard. Nous voudrions bien le croire, mais ... . Examinons d'abord de plus près ce chapitre.

Selon les dires de l'ermite, il n'y a donc point de hasard. Zadig est plus ou moins convaincu par les deux premières illustrations de l'ermite, mais il se révolte contre le troisième exemple. Il proteste contre l'apparente cruauté de l'ermite qui soutient que ses crimes ont prévenu de plus grands malheurs. Zadig objecte trois fois et refuse d'accepter aussi bien la nécessité que l'apparente universalité et l'injustice du mal. L'ermite répond mal, avec des arguments circulaires: le mal existe partout sauf au Paradis; comme la terre n'est pas le Paradis, il y a nécessairement du mal ... . A la dernière intervention de Zadig, l'ermite devenu ange et équipé d'ailes, s'envole. Zadig s'agenouille, adore et proteste en prononçant un dernier 'mais' (p. 57). L'ange dit ensuite à Zadig d'aller à Babylone. Zadig ne suit pas ce conseil et marche plutôt au hasard. En dépit du hasard, ou peut-être à cause de lui, il arrive à Babylone. Voltaire le rusé joue avec son lecteur, il fait même de notre dilemme un jeu de mots!

Chez Voltaire, le mot 'adorer' endosse, Van den Heuvel nous le signale, une signification ironique, et l'envolée de l'ange nous fait penser au derviche dans *Candide*, qui fermera la porte au nez du héros interrogateur; elle nous fait penser également au Sirien dans *Micromégas* qui s'envole, lui aussi et laisse un beau livre illisible pour l'Académie des Sciences.[7] Cependant Zadig s'agenouille et finit par entrer dans la ville de Babylone. Peut-être qu'il accepte, en murmurant, la notion de destin. On peut aussi croire que Zadig agenouillé n'est ni debout ni aplati, les bras étendus en forme de croix (comme ses ancêtres médiévaux l'étaient), et qu'il n'accepte ni ne rejette le destin, qu'il est en proie à des doutes. Nous ne trouvons la clef au reste du conte ni dans ce chapitre ni à la fin où Zadig pense au grain de sable (il doute) et où il adore et bénit la Providence dans une seule phrase qui résout tout le conte et rien.

Si l'on considère le destin comme le thème central du conte, il y a beaucoup d'exemples d'hommes qui échappent à leur destin, notamment le héros, Zadig, qui devait mourir au moins trois fois, mais qui s'est échappé grâce à un mauvais médecin, un ami et un billet doux livré très à propos, entre la poire et le poison, pour ainsi dire. Zadig intervient et change le destin du pêcheur, et même le sien propre. Dans beaucoup de chapitres on ne parle nullement du destin, mais plutôt de l'infidélité des femmes, des hommes, de l'inaptitude des médecins, de la corruption des juges, etc.

Le destin et le hasard sont, certes, des thèmes importants dans *Zadig*, mais sont-ils des thèmes centraux? Voltaire lui-même ne nous prouve-t-il pas que le hasard à la fois existe et n'existe pas? Si l'on divise le conte d'après le thème du hasard, les chapitres se groupent en quatre catégories: 1) le destin ne rémunère pas la bonté; 2) le destin rémunère la bonté; 3) le destin rémunère le mal; 4) le destin peut rémunérer la bonté. Ces groupes s'opposent symétriquement et se réduisent par là à rien en termes de signification. La structure du conte, de ce point de vue, ressemble au discours de Zadig qui dit une chose ('mais') et en fait une autre (il s'agenouille), et à celui du narrateur qui le laisse s'égarer, mais dans la direction de Babylone. Si l'on poursuit ce thème on devra conclure comme le fait le narrateur de l'*Histoire du Bon Bramin:* 'il y a là de quoi parler beaucoup' (p. 116). Cette interprétation ressemble à celles de René Pomeau et de Lester Crocker. Selon eux, Voltaire nous explique un dilemme, mais n'y voit pas de solution lui-même.[8] Nous ne voudrions cependant pas accepter un 'non liquet' comme étant la principale signification de notre conte.

Aussi avons-nous pensé à donner à *Zadig* un sous-titre: 'ou la recherche du bonheur.' Zadig commence sa quête d'une manière assez systématique: il cherche le bonheur chez les femmes (nobles et citoyennes), à la campagne, à la cour, dans les sciences, chez les savants et les honnêtes gens. Mais après le quatrième chapitre il s'arrête. Il oublie apparemment qu'il cherche le bonheur — il se croit d'abord heureux et est ensuite obligé de chercher Astarté. La quête d'Astarté remplace ainsi la quête du bonheur. Astarté représente non seulement la passion mais aussi le pouvoir. Sans elle, Zadig se croit le plus malheureux des hommes — mais il se trompe, et nous le savons. D'une part, le pêcheur semble bien plus à plaindre que Zadig, même si celui-ci nous assure que les malheurs du coeur sont les plus graves. D'autre part, le narrateur nous a déjà appris que la lune de l'absinthe suit la lune de miel. Le bonheur est instable, indéfinissable et trompeur — il ne faut pas croire aux apparences.

Nous avons alors imaginé un deuxième sous-titre dans l'espoir de découvrir le lien secret entre tous les chapitres du conte: 'ou l'amitié.' En effet, le thème de l'amitié parcourt l'ensemble du conte. D'abord, Zadig a des amis uniquement à cause de ses richesses. Ensuite, il paie Cador pour s'assurer de son amitié. Plus tard il envoie le pêcheur chez Cador, car il semble avoir enfin appris que l'amitié dépasse les considérations monétaires. Et Zadig serait à la fin le seul monarque sur terre qui eût un ami.

Mais Zadig semble toujours préférer la passion à l'amitié, et pourtant il est continuellement sauvé par ses amis. S'il n'est pas victime du destin, c'est grâce à ses amis qui arrivent à en contrecarrer les coups. Zadig rouspète sans cesse contre son mauvais destin, mais lorsque ses amis le tirent d'affaire, il loue la Providence. Il ne se montre que rarement reconnaissant. Cador ressemble à un sage, peut-être à Tiberge, l'ami de Des Grieux dans *Manon Lescaut* de Prévost. Il donne de bons conseils à Zadig qui ne les suit pas. Il avertit Zadig du danger et ce dernier refuse d'y faire attention. Cador le sauve: des prêtres, des juges, des envieux, des jaloux. Il est capable de sauver Zadig de tout sauf de la passion qui dévore notre héros. (Et il ne faut pas oublier que le narrateur condamne cette passion et la traite comme une maladie: 'Zadig fut longtemps malade, mais enfin la raison l'emporta sur son affliction...' [p. 4].) Au début Zadig a pour Sémire un attachement solide et vertueux tandis qu'elle l'aime avec passion. La passion se montre fragile et ne dure qu'aussi longtemps que Zadig garde sa beauté physique. Par la suite, presque toutes ses mésaventures sont causées soit par la passion des autres et souvent celle des femmes (comme celle de l'envieux), soit par sa propre passion qui le rend aveugle — qui ne lui permet pas de raisonner ni de reconnaître ses amis. Zadig, en ami raisonnable, sauve la vie à Almona qui est en proie à la passion du paraître. A son tour, Zadig est sauvé par Almona et pour la première fois il montre de la reconnaissance — il se jette à ses pieds. Il aide ensuite le malheureux laitier devenu pêcheur en lui disant d'aller voir Cador. Zadig reconnaît, du moins indirectement, l'amitié et le fait qu'il pourra toujours compter sur Cador. Et, détail intéressant à noter, quand le pêcheur veut montrer sa reconnaissance, il se jette aux pieds de 'l'ami de Cador' (p. 41). Le narrateur, pour la première fois, désigne Zadig autrement que par son nom. Zadig est enfin devenu ami et, deux chapitres plus loin, il voit clair et distribue sagement des bénéfices à ses amis. Zadig endosse finalement les caractéristiques de son ami raisonnable. Mais cette fin est un tant soit peu féerique car elle combine pouvoir, amour et amitié, mélange instable, sinon impossible. D'ailleurs, si le thème du conte est vraiment l'amitié, plusieurs chapitres tels que 'L'Ermite,' 'Le Souper' et 'La Femme battue,' sont superflus. L'amitié doit alors être considérée comme un sous-thème, un leitmotif, et nous devons de nouveau nous mettre à la recherche du thème principal de *Zadig*.

Voltaire a fait, au début du conte, des promesses auxquelles notre naïveté nous permettra de croire. Il nous a averti de ce que son conte est 'un ouvrage qui dit plus qu'il ne semble dire' (p. 1). En comparant *Zadig* aux histoires des *Mille et une Nuits*, le sage Oulong note que ces dernières 'sont sans raison et ne signifient rien' (p. 2), ce qui permet d'inférer que *Zadig* a une raison et signifie quelque chose de plus qu'il ne semble signifier à première vue.

En cherchant une signification qui donnerait de l'unité au conte, nous

nous sommes rendu compte du fait que tout est contredit ou contradiction dans *Zadig*. Si Zadig se jette à genoux devant l'ermite devenu ange, il se prosterne aussi devant les chandelles avec lesquelles il illustre le ridicule de l'adoration. Si Zadig appelle l'ermite 'un ange sauveur,' le pêcheur donne ce même titre à Zadig. Si Zadig est victime des passions coupables de sa femme, il inspire ces mêmes sentiments à la femme d'un autre. S'il est rendu malheureux par l'infidélité des êtres humains, il est aussi sauvé par cette même infidélité. Zadig est esclave et roi, victime et vainqueur. Chaque épisode a un parallèle qui semble donner une leçon contradictoire. Chaque geste et chaque symbole est suivi de son antithèse. Souvent ces antithèses sont introduites par un 'mais.' Le mot-clé du chapitre de l'ermite est répété dix fois dans ce seul chapitre et au moins cinquante-sept fois dans le reste du récit.[9] 'Mais' présente, par définition, une contradiction, et Voltaire s'en sert pour créer des contradictions même là où l'on ne s'attend pas à en trouver. Il emploie l'antithèse synonymique, procédé que Deloffre[10] a relevé chez Marivaux. Dans notre texte nous trouvons les exemples suivants: 'Zadig n'était point un assassin ... mais il était coupable du sang d'un homme' (p. 26); 'je n'étais point coupable mais ... je vais mourir criminelle' (p. 22). Ces 'mais' ajoutent à l'effet comique créé par le caractère inattendu de la phrase qui suit.[11] Dans l'exemple suivant nous nous attendons, après la boule de neige linguistique de toutes les possibilités d'agir, à une action: 'Zadig fut tenté à la fois d'éclater de rire, de dire des injures au révérend père, de le battre, et de s'enfuir mais ... il ne fit rien de tout cela' (p. 55). *Zadig* est parsemé de 'mais,' mot qui devient, avec la litanie des malheurs de Zadig, un refrain dont l'ironie a déjà été signalée dans les études de Topazio et Van den Heuvel. Ceux-ci citent l'emploi du 'mais' dans d'autres textes de Voltaire tels que la lettre à Madame Denis: Voltaire lui y écrit que les soupers du roi Frédéric II sont délicieux ... mais. Le 'mais' clot les trois débats (ou mini-débats) philosophiques de *Zadig:* le souper, l'esclavage et l'ermite. Les trois conclusions se contredisent mutuellement et contredisent au moins une partie du débat à l'intérieur de la discussion. Un véritable crescendo de 'mais' accompagne le thème secondaire du bonheur instable.

Après seulement quatre chapitres, Zadig croit à la possibilité du bonheur. Le narrateur intervient pour nous informer que Zadig se trompe. Quelques chapitres plus loin, Zadig pense qu'il est enfin heureux, mais il commence lui-même à se douter de l'instabilité du bonheur. Il demande alors au perroquet du roi si le bonheur sera de courte durée. La question est posée d'une manière qui indique le pessimisme naissant de Zadig. L'oiseau répond affirmativement et le prochain malheur s'annonce déjà. Enfin, Zadig se dit, en songe, bien entendu (il parle aux oiseaux, mais il ne se parle pas), que son bonheur va bientôt s'évanouir. Le narrateur, l'oiseau et Zadig ont tous eu raison, chacun à son tour. Alors, quand à la fin du conte nous apprenons que Zadig fut roi et fut heureux, la phrase nous choque par sa qualité insolite. Après dix-neuf leçons et une centaine de pages d'exemples, le lecteur a envie d'ajouter: 'mais où est le serpent?' Et il y a beaucoup de serpents potentiels. Zadig est heureux grâce à sa femme. Mais il a déjà été trompé par ses deux premières épouses. Nous avons appris que la lune de miel est suivie de la lune de l'absinthe. Nous nous souvenons du mariage de Sétoc: 'Sétoc devait aller ... dans l'île de Serendib; mais le premier mois de son mariage qui est, comme on le sait, la lune de miel, ne lui permettait ni de quitter sa femme, ni de croire qu'il pût jamais la quitter' (p. 60). Nous nous rappelons Zadig qui a répudié Azora quelque temps après le second mois (p. 6). Et nous avons sous les yeux le sage exemple du pêcheur qui, lui, prend l'argent et refuse la femme (p. 59). Même dans l'exemple de Sétoc qui n'a pas encore quitté sa femme, la phrase est négative et suggère que le bonheur ne va pas durer. Il est donc naturel de penser à la lune de l'absinthe quand enfin Zadig épouse Astarté. Voltaire décrit cette dernière scène avec des superlatifs: 'ce fut le plus beau siècle de la terre' (p. 59). Nous avons déjà assez souvent entendu que Zadig fut 'le plus malheureux des hommes' (p. 41, par exemple). Si nous rejetons la première exagération, nous en ferons autant des autres. Nous ne croirons pas au véritable bonheur de Zadig et nous ajouterons nous-mêmes la phrase prophétique: 'mais il se trompe.'

Tout est donc contradictoire et les apparences sont trompeuses. Le conte, selon Voltaire, dit plus qu'il ne semble dire. Les synonymes se transforment en antonymes. Le grand débat dans le chapitre du 'Souper' n'est pas un débat du tout, car Zadig, même s'il n'a pas tout compris, a assez compris pour savoir que tous les interlocuteurs partagent le même avis. Zadig montre à Sétoc le ridicule de ses croyances en l'imitant avec exagération. L'ange montre à Zadig que ce qui paraît bon peut être mauvais et vice versa.

Zadig attribue constamment ses malheurs à des raisons superficielles. Il croit, par exemple, avoir perdu la reine et le royaume parce qu'il a fait la grasse matinée et non parce qu'il a des ennemis motivés par l'envie et la jalousie.

Tout le mal qui arrive à Zadig est causé par les passions des hommes. Ce n'est pas le destin qui le condamne à mort, mais les hommes — les hommes avec leur envie de l'argent et du pouvoir. Zadig est incapable de comprendre les femmes qu'il juge par leur beauté et par la classe sociale. Les femmes ne sont guère meilleures que les hommes et les quittent quand ils deviennent laids ou sentent l'approche de la mort. Les femmes, tout comme les hommes, sont elles aussi influencées par le paraître. Zadig est un habile juge qui sait résoudre les problèmes des autres, mais il est incapable d'identifier ses propres ennemis. Quand on loue son sens de la justice, il ne comprend pas que l'on le flatte et que l'on ne l'aime que pour sa position dans le gouvernement. Même la veuve Almona est victime des apparences: elle détestait son mari mais elle se brûlera pour sauver les apparences: 'Tout le monde se moquerait de moi si je ne me brûlais pas' (p. 30). Et les juges corrompus sont condamnés moins pour avoir aimé cette femme que pour avoir été hypocrites.

Les méchants semblent heureux et les heureux, méchants. Le malheur même est mis en question. Nous assistons à un concours de malheurs entre Zadig et le pêcheur, concours à la fin duquel Zadig se déclare le gagnant car son malheur invisible est pire qu'un malheur visible. Zadig gagne le combat contre Itobad, mais il n'est pas reconnu vainqueur à cause, encore une fois, des apparences. Itobad revêt l'armure blanche de Zadig et les juges et le public vont couronner non le vainqueur mais ses vêtements. Ils permettent à Zadig de se présenter pour expliquer les énigmes, non à cause de sa probité, mais à cause de sa réputation de probité. Astarté n'est pas simplement heureuse d'être réunie avec Zadig, mais elle 'goûtait ... la douceur de voir son amant digne *aux yeux de l'univers* d'être son époux' (p. 59). Zadig devient roi, époux et heureux. Il règle ses dettes, punit ses ennemis, et le lecteur attend la prochaine mésaventure car il a bien appris qu'il ne faut pas juger quoi que ce soit, que tout est trompeur, que tout est instable.

Une comparaison de *Zadig* et de *Candide* révèle une différence d'ordre structural. Candide n'est jamais seul. Il est toujours accompagné de Pangloss ou de Martin. Il y a donc possibilité de dialogue. Candide peut essuyer des mésaventures et murmurer contre son destin tout en écoutant une leçon d'optimisme. Il existe une certaine tension entre les personnages et Candide a le choix d'apprendre la leçon qu'il tirera des événements ou des dires de ses amis. Zadig, au contraire, est seul et doit agir et énoncer la leçon en même temps. Il n'est pas parfaitement sage, ni au début ni à la fin. Il est cependant obligé de nous résumer sa propre leçon avant de l'avoir apprise. Il ne peut certainement pas nous dire ce qu'il ne sait pas, et Voltaire ne peut pas faire intervenir trop souvent des oiseaux parlants ou faire parler notre héros en songe. Or, c'est en faisant de Zadig un mauvais élève que Voltaire pourra exprimer ses idées. Zadig dira le contraire de ce que Voltaire veut dire, ou il se taira. Voltaire nous signale que tel est le rôle des 'mais.' Rien n'est ce qu'il semble être.

Le conte est doublement ironique, car Zadig répète une litanie de malheurs qui est instable. Les malheurs inscrits sur la liste changent continuellement. Seule reste constante l'ironie de la contradiction: Zadig, le plus intelligent des hommes, tire toujours la mauvaise conclusion des leçons et fait des erreurs de logique aussi évidentes que l'est l'ironie.

Voltaire n'a pas inventé l'histoire du nez; il n'a pas inventé l'histoire de l'ermite non plus. Ce commentaire n'a rien d'original: Fréron accusa déjà Voltaire de plagiat il y a deux siècles. Parmi les sources possibles figurent de vieilles légendes hébraïques, le Coran, les *Vitae patrum*, les *Gesta Romanorum*, le *Libro de los enxemplos*, les *Exempla* de Jacques de Vitry, *De causa dei* de Bradwardine, Luther, Antoinette Bourignon, Herbert, Jean le Jeune, Henry More, Cotton Mather et le poème *The Hermit* de Parnell.[12] L'histoire de l'ermite est un renversement complet du texte médiéval français, texte que Voltaire a sans doute connu et utilisé, comme il a utilisé celui de Parnell.

Le texte médiéval a pour but de calmer les esprits d'un ermite qui a commencé à douter d'une Providence qui semble punir les bons et enrichir les méchants. La leçon que l'ange lui donne est celle de la résignation à la Providence, car l'apparente injustice terrestre sera corrigée dans une autre vie où règnera la grâce. Les exemples, aussi illogiques qu'ils semblent, servent à rassurer l'ermite qui se tait devant le discours de l'ange et qui retourne humblement à sa cave pour prêcher la gloire de Dieu. Le conte illustre l'idée exprimée par Saint Thomas d'Aquin et selon laquelle les actes de violence inexplicables sont en eux-mêmes la démonstration de l'existence de Dieu. La violence de l'ange sauve l'âme du pêcheur en puissance, et la vie de la victime potentielle n'est qu'un dérivé accidentel.

Dans la version de Voltaire les mêmes épisodes qui calment le doute de l'ermite inspirent le doute à Zadig. Voltaire remplace le règne de la grâce par le règne de la justice. Il n'est plus question d'une vie dans l'Au-delà. Tout se passe sur terre — la gloire est terrestre. Il est vrai que Zadig ne peut pas lire dans le livre des destins, mais il sait résoudre des énigmes. Dans le chapitre 'Le Brigand,' Voltaire présente l'idée qui choque tant Zadig qu'il ne dort pas de la nuit: 'L'empire est déchiré, et ce brigand est heureux: ô fortune! ô destinée! un voleur est heureux et ce que la nature a fait de plus aimable a péri peut-être d'une manière affreuse' (p. 38). Voltaire reproduit ensuite une histoire qui 'prouve' que la justice existe après la mort, et qui met en doute le rôle de la Providence: Zadig parle et l'ange, muet, s'envole. Dans la version médiévale, c'est l'ange qui parle et l'être humain ne dit mot. En réalisant les rêves de Zadig à la fin du conte, Voltaire est très ironique: ou bien il veut nous faire croire à un autre exemple de l'illogique du système du bien et du mal (pourquoi faire gagner Zadig maintenant?), ou bien la leçon du conte est une négation de l'éternel qui est ainsi transitoire et terrestre.

Dans *Zadig* les méchants paraissent heureux et souvent innocents, et les bons, malheureux et coupables. Toute récompense est uniquement terrestre et limitée au présent. Les récompenses ne sont cependant ni visibles ni compréhensibles. C'est une vue extrêmement pessimiste. Voltaire n'offre aucun remède. L'homme est entièrement incapable de juger, car comment juger autrement que selon les apparences? Zadig aide la femme battue une première fois car elle semble souffrir, mais il apprend qu'elle aime être battue. La deuxième fois Zadig refuse de l'aider car il ne croit plus aux apparences. Nous ne saurons jamais s'il avait raison, mais habitués à l'ironie, nous sommes portés à croire qu'elle souffrait véritablement la deuxième fois.[13]

Zadig est composé d'un nombre d'histoires dont trois sont puisées dans l'histoire littéraire et intercalées dans le reste du récit: l'histoire du nez, celle du bûcher et celle du pêcheur. Chacune de ces histoires est précédée d'une introduction (ou pré-histoire) ou suivie d'une conclusion (ou post-histoire) qui souligne la leçon de l'histoire empruntée. L'histoire du 'Borgne' précède celle du 'Nez.' Dans les deux cas, Zadig faillit perdre une partie de son anatomie, et perd soit sa femme soit sa fiancée. Dans 'Le Borgne,' la femme qui semble aimer n'aime que les apparences. Elle n'aimera pas un Zadig borgne. Dans 'Le Nez,' Zadig essaie d'apprendre à cette femme de ne pas juger selon les apparences. Il se trompe et apprend, à son malheur, que sa femme ne l'aime pas. Zadig lui aussi juge selon les apparences et attribue le mal d'abord à la classe sociale de sa fiancée et ensuite à toutes les femmes sans penser que le mal vient de la nature passionnée de sa fiancée dans la première histoire et de sa propre passion dans la deuxième. Azora est raisonnable et logique. Zadig, mort, n'aura vraisemblablement pas besoin de son nez. Zadig, motivé par la vengeance, avait construit le piège dans lequel il a fini par attraper sa femme et perdre, par conséquent, son propre bonheur.

Nous apprenons dans 'Le Bûcher' et 'Les Rendez-vous' que la société juge du paraître plutôt que de l'être et que le résultat est le règne de l'injustice. 'L'Ermite' et 'Le Pêcheur' nous enseignent qu'il est impossible de croire à un système divin de récompenses et de punitions, car les récompenses et les punitions sont indistinguibles. Les personnages victimes des événements ne savent même pas s'ils sont vraiment malheureux.

La gradation que nous observons dans les trois paires d'histoires va de l'individu à la société au surnaturel. Personne ne sait distinguer l'être du paraître. Selon la leçon qui unit ces histoires, il faut examiner d'un œil critique l'image que l'on projette autant que celle que l'on perçoit.

Le 'mais' constitue l'objection rationnelle à la démonstration surnaturelle. C'est le doute qui est inspiré par ce qui était à une époque révolue l'évidence irréfutable de la Providence. En général, le 'mais' constitue un leitmotif qui sert à nous avertir de ce que le terrain est instable et changera d'aspect avec notre point de vue.

Ainsi averti, le lecteur a la satisfaction d'être mieux informé que le public envisagé dans le texte. Il ne croit pas qu'Itobad ait vaincu Zadig. Il ne pense pas, comme le pêcheur, que Zadig soit un ange sauveur, et il ne sera certainement pas coupable d'avoir jugé quelqu'un par la couleur de ses jarretières.

Il y a donc complicité entre le narrateur et son lecteur. Cette complicité est exprimée sous forme de dialogue. Le narrateur communique son opinion dans les brefs commentaires qu'il ajoute à la narration. En voici un exemple: 'Zadig disait: "Je suis donc enfin heureux!" Mais il se trompait' (p. 15). L'opinion tranchante du narrateur est communiquée dans un style nu et presque télégraphique. L'oiseau, lui aussi, transmet l'opinion du narrateur et Zadig la réitère en rêvant. Le 'mais' sert également de rappel constant à la réalité du narrateur. Le lecteur ne doit rien accepter comme acquis. Des événements au vocabulaire, tout se contredit. Voltaire emploie le même procédé dans tous ses contes. Le lecteur est systématiquement éduqué à raisonner et à ne rien croire. Les morts se raniment. Zadig guérit en dépit des médecins. Il est injustement condamné à l'esclavage par les Egyptiens qui 'étaient alors justes et humains' (p. 26). Il est injustement condamné à mort par un bon roi qui 'crut tout ce qu'il voyait, et imagina tout ce qu'il ne voyait point' (p. 21). Même les descriptions sont trompeuses. Comme l'a noté Jean Sareil, Almona avait un nez 'qui n'était pas comme la tour du mont Liban' (p. 35), c'est-à-dire, qui était probablement petit.[14]

Le narrateur communique avec son lecteur, et lorsque celui-ci s'arrête et relit ce texte stylistiquement trompeur, le dialogue s'installe. Il y a aussi dialogue lorsque le lecteur lit, 'Zadig fut roi, et fut heureux' (p. 59), et substitue le mot 'mais.' Il y a enfin dialogue quand le lecteur, arrivé à la dernière page, est prêt à continuer et tourne la page dans la ferme conviction que le règne de la gloire, de l'abondance, de la justice et de l'amour sera suivi d'une nouvelle mésaventure. Le narrateur est étrangement absent dans le dernier paragraphe. Il n'y a pas un seul 'mais.' Le nain semble avoir renoncé à la peinture et le perroquet semble avoir été oublié. Et Zadig, qui ne prononce pas un seul mot, semble avoir bien retenu ses leçons. Mais quelles leçons? Celle de l'ange (on ne peut pas échapper à sa destinée) et celle du brigand (on peut changer son destin par ses propres efforts). Si le lecteur n'a pas déjà dit, 'mais quel sera le serpent,' il fera comme Zadig devant l'ermite: il demandera 'la permission de parler' (p. 56) et il objectera, comme Zadig: 'Mais ... .' 'Mais ... s'il n'y avait que du bien et point de mal' (p. 56)? L'ermite a déjà répondu: 'Alors, ... cette terre serait une autre terre' (p. 56). Voltaire en donne l'illustration concrète avec cette fin féerique. De plus, un subtile changement s'opère. A la place d'un Zadig qui demande une réponse à l'ermite devenu ange, Voltaire substitue le lecteur. Celui-ci pose des questions au narrateur qui s'absente tout comme l'ange s'envole. Il est ironique que ces questions sont celles que le narrateur a déjà posées au lecteur et les mêmes que Zadig a posées à l'ange. Le dialogue prend donc la forme d'un cercle parfait qui finit là où il a débuté. Voltaire nous laisse une question plutôt qu'une réponse. *Zadig* n'est pas l'histoire de la destinée. Le titre est trompeur. Zadig n'est sauvé ni par l'ange, ni par un hasard miraculeux, mais par ses amis et ses amies raisonnables. *Zadig* est l'histoire de l'être et du paraître. Le conte dit bien 'plus qu'il ne semble dire' (p. 1). Ce n'est pas un simple conte de fées; presque chaque phrase est chargée d'une double signification. Mais *Zadig* dit aussi moins qu'il ne semble dire, parce que le double sens est parfois faux sens, et que, lorsqu'il s'agit du hasard, il ne peut y avoir de réponse.

ROSEANN RUNTE
Université Sainte-Anne

Notes

1 Jacques Van den Heuvel, *Voltaire dans ses contes* (Paris: Armand Colin, 1967), pp. 170-79. Toute référence ultérieure à cette étude sera notée entre parenthèses dans le texte.

2 Roland Barthes, ed., *Romans et contes de Voltaire* (Paris, s.d.), p. 45.

3 François-Marie Arouet de Voltaire, 'Zadig,' *Romans et contes,* ed. H. Bénac (Paris: Garnier, 1960), p. 3. Toute référence ultérieure à cette édition sera notée dans le texte.

4 Voir, par exemple, Frederick M. Keener, '*Candide:* Structure and Motivation,' *Studies in Eighteenth-Century Culture,* 9 (1979), 405-28; Roy S. Wolper, '*Zadig,* a Grim Comedy?' *Romanic Review,* 64, 4 (novembre 1974), 237-48.

5 Nancy Senior, 'The Structure of *Zadig,*' *Studies on Voltaire and the Eighteenth Century,* 135 (1975), 135-41. Dans son introduction, Senior résume les arguments de V. Mylne, F.C. Green, N. Torrey et L. Crocker.

6 Voir, par exemple, E.J.H. Greene, 'The Destiny of *Zadig,*' *Esprit Créateur,* 3, 4 (hiver 1967), 243-51, et June Sigler Siegal, 'Voltaire, *Zadig,* and the Problem of Evil,' *Romanic Review,* 50 (1959), 29-32.

7 Voir George A. Perla, 'Zadig, Hero of the Absurd,' *Studies on Voltaire and the Eighteenth Century,* 143 (1975), 56-66; voir également Van den Heuvel, p. 172.

8 René Pomeau, *La Religion de Voltaire* (Paris, 1956), p. 244; Lester G. Crocker, *An Age of Crisis: Man and World in Eighteenth-Century French Thought* (Baltimore: Johns Hopkins University Press, 1959), pp. 63-67.

9 Voir Virgil W. Topazio, *Voltaire: A Critical Study of His Major Works* (New York: Random House, 1967), pp. 34-35, pour une brève étude du rôle du 'mais' dans d'autres œuvres de Voltaire; voir aussi la communication non publiée de Timothy Scanlan, 'Le "Mais" dans *Candide,' Atlantic Society for Eighteenth-Century Studies Conference, Université de Moncton, 1978.*

10 Frédéric Deloffre, *Une Préciosité nouvelle: Marivaux ou le marivaudage* (Paris: Armand Colin, 1967), passim.

11 Henri Bergson, 'Laughter,' dans *Comedy,* ed. Wylie Sypher (Garden City: Doubleday, 1956), p. 94.

12 Nous avons examiné ces sources dans une communication donnée au Congrès des Sociétés Savantes à Edmonton en 1981.

13 Il n'y a pas de preuves textuelles de sa souffrance. Cependant, lorsque je pose cette question à mes étudiants et à mes collègues, ils me répondent tous sans hésitation que Zadig se trompe comme d'habitude et qu'il devrait s'arrêter pour aider la deuxième demoiselle véritablement en détresse.

14 Jean Sareil, 'Le Comique par non-sens et faux sens dans les contes de Voltaire,' *Studies in Eighteenth-Century Culture,* 9 (1979), 482.

5. Textes et réécritures de textes: le cas des *Fêtes grecques et romaines* de Louis Fuzelier

Le Ballet nouveau des *Fêtes Grecques & Romaines* a été représenté pour la première fois sur le Théâtre de l'Opéra le Mardy, 13 Juillet.[1]

C'est ainsi que *Le Mercure* de juillet 1723 annonça la première représentation du ballet héroïque en trois actes et avec prologue du librettiste Louis Fuzelier. En août de la même année les représentations continuaient, mais non sans quelques changements apportés au spectacle:

Mlee [sic] Prévost a produit une éleve sur le Theatre de l'Opera qui a débuté le Jeudy 26 Aoust dans le troisiéme Acte des *Fêtes Grecques & Romaines;* ... On la nomme Mlle Richalet; elle n'a pas encore quinze ans.

L'introduction de cette jeune danseuse dut modifier — ne serait-ce que de façon minime — l'exécution du ballet; le texte écrit fut modifié également, si l'on en croit la suite de l'article du *Mercure* qui cite une chanson écrite en l'honneur de Mlle Richalet.[2]

La mise en espace par laquelle un texte se métamorphose en voix et en mouvements est un processus propre au théâtre, ainsi que l'est celui par lequel des représentations successives évoluent, jusqu'à entraîner parfois la réécriture du script de départ. Il s'agit d'un *passage* entre deux 'textes': entre un texte écrit et une réalisation scénique que les théâtrologues dénomment 'texte' aussi; entre telle réalisation et une modification

0824-3298/84/0300-0077 $01.25/©C.S.E.C.S./S.C.E.D.S.

ultérieure; enfin, entre les modifications initiées par des acteurs et un texte d'auteur dûment réécrit pour les inscrire. La fréquence des 'passages entre deux textes' occasionnés par *Les Fêtes grecques et romaines* fut particulièrement élevée, surtout si l'on considère que Fuzelier lui-même fit chaque réécriture. Scénariste, parolier, auteur de préface, co-rédacteur du *Mercure* et probablement critique de son propre ballet, parodiste curieusement acharné contre lui-même, Fuzelier retravailla son sujet non moins de huit fois. *Les Fêtes grecques* passèrent par deux ébauches, une édition précédée d'une préface critique, un compte-rendu journalistique faisant état d'une version antérieure du ballet, une parodie non-imprimée et non-représentée intitulée *La Rencontre des opéra,* une deuxième parodie intitulée *Les Saturnales,* et une version abrégée de la deuxième parodie intitulée *Le Debris des Saturnales.*

Ce mouvement continu d'une forme à une autre caractérisait donc la pratique théâtrale de Louis Fuzelier. Relevant d'une esthétique du flux et de l'indécis, cette pratique préférait le dynamisme au statisme. Là où, par exemple, une pièce de Voltaire aboutissait presque invariablement à l'édition, autorisant un René Pomeau à affirmer que 'L'imprimé constitue au dix-huitième siècle la forme éminente de la communication,'[3] une pièce de Fuzelier tendait à générer une succession de textes plus ou moins éphémères. On sait que de nombreuses pièces de Fuzelier furent imprimées: ses livrets d'opéra parurent dans la série des oeuvres représentées à l'Académie Royale de Musique et publiées par la famille Ballard; on lui connaît une quarantaine de pièces (écrites par lui seul ou en collaboration) dans *Le Théâtre de la foire, ou l'Opéra Comique,*[4] ainsi que plusieurs textes publiés séparément. Il reste cependant une grande part de sa production théâtrale de quelque 237 pièces qui se caractérise, soit par des versions multiples, soit par des versions manuscrites et souvent fragmentaires.[5] Le texte de son opéra le plus célèbre, *Les Indes galantes* (1735), fut augmenté par l'entrée intitulée 'Les Sauvages.' Le très populaire *Scaramouche pédant,* pantomime 'publiée' en 1711, fut la quatrième de cinq versions plus ou moins différentes et dont l'existence est attestée par des procès-verbaux ou des manuscrits inédits.

Or, l'intérêt des œuvres flottantes, 'imparfaites,' ou inachevées se reconnaît de plus en plus depuis que Jean Bellemin-Noël ouvrit il y a dix ans déjà la notion de 'texte' aux *presque-textes* et aux *avant-textes:*

> Il ... faut ... tâcher d'intégrer à la recherche littéraire tout ce à quoi on voudra conférer le *statut de texte,* à commencer par les *presque-textes* que sont les œuvres de littératures dites "marginales" (orales, populaires) — cela est en train de se faire — et en continuant par les *avant-textes,* où qu'on les trouve et quelque texte qu'ils préparent.[6]

Sensible aux périls d'un délire terminologique, j'ajouterais néanmoins *après-texte* à cette liste pour décrire le produit des réécritures successives que furent les innombrables parodies de Fuzelier et de bon nombre de ses contemporains.

Singulier par son ampleur, le foisonnement de textes associé au cas des *Fêtes grecques* est pourtant typique d'une *autre* 'forme éminente de la communication' au dix-huitième siècle, le théâtre.[7] Mais comment capter ce dynamisme? Le corpus peu connu du théâtre de Fuzelier, et en particulier la série de textes écrits autour des *Fêtes grecques*, nous permet d'en isoler certaines des composantes fondamentales. Cette série constitue une trace importante d'une pratique théâtrale communiquant le théâtre à la fois comme texte achevé, comme lieu et objet de critique, et comme processus de production. C'est cet aspect de mouvance que je veux poursuivre – impossible de l'arrêter de façon définitive – dans les propos que voici.

Parmi les papiers personnels de Fuzelier se trouve une feuille avec l'en-tête 'Pour le Roy/ les festes Romaines ou de julie. .. .'[8] Ce document où fourmillent les ratures est à peine une ébauche; les sujets dispersés qui formeront les trois actes du ballet des *Fêtes grecques*, côtoient une variété d'autres sujets, abandonnés ou utilisés ailleurs.[9] L'amas de motifs gravite autour du mot 'feste,'[10] contient quelques vers rédigés d'un 'prologue' évoquant une participation active du roi,[11] et renvoie à des recherches ultérieures.[12] *Avant-texte* ouvert, certes, mais qui délimite toutefois les possibilités d'un programme de divertissement royal à venir.

Dans le même dossier se trouve une seconde feuille contenant le résultat des recherches de Fuzelier sur certaines fêtes de l'antiquité,[13] ainsi que le plan sommaire de son ballet.[14] D'aspect plus soigné que la première feuille, ce document semble avoir été façonné par les exigences d'une mise en spectacle. Aussi rencontre-t-on plus souvent le mot 'jeux,' qui théâtralise 'fêtes,' et qui suggère une préoccupation avec le mouvement ('jeux' au sens de 'concours,' 'exercices militaires,' 'combats,' 'assauts') et avec les problèmes de la concrétisation ('on y representoit,' 'se representoient,' 'jeux sceniques,' et 'on peut mettre en action...'). Si la première feuille laisse entrevoir l'immensité du fonds culturel dans lequel Fuzelier puisera les motifs de son programme, la seconde révèle une structuration de ce fonds sous l'égide du code d'une éventuelle représentation.

Conçues pour être jouées dans la salle des machines au Palais des Tuileries,[15] les *Fêtes grecques* finirent par l'être devant un public 'ordinaire' au Théâtre de l'Opéra. On put suivre le spectacle grâce à un livret imprimé.[16] Fait relativement exceptionnel pour l'opéra, ce texte était précédé d'un 'avertissement' de quatre pages où Fuzelier anticipait les critiques qu'on pourrait éventuellement lui adresser. Contraint de "recycler" son texte, il s'efforçait sans doute d'en justifier l'étrangeté en faisant appel au goût de la nouveauté.[17] Il se félicitait en outre d'avoir exploité l'histoire plutôt qu'une thématique fictive.[18] La singularité de cette incursion de l'historique dans un genre voué au merveilleux sera constamment reprise dans les textes ultérieurs de Fuzelier.

Annoncée dans la préface, la portée novatrice des *Fêtes grecques* est de nouveau proclamée dans un prologue auto-réflexif qui prépare l'action théâtrale. Clio, Muse de l'histoire, fait savoir à Erato, Muse de la musique, que la vérité a aussi droit de cité dans l'art de l'opéra.[19] Apollon, qui préside la rencontre des muses, annonce que les jeux d'Erato ont aussi besoin de l'appui de la danse, figurée par Terpsicore.[20]

Le ballet lui-même met en scène trois couples d'amoureux célèbres dont les tête-à-tête ouvrent la voie à trois fêtes dansées. Le premier acte, 'Les Jeux olympiques,' commence par l'entretien d'Alcibiade et Timée, le deuxième, 'Les Baccanales,' par celui de Marc-Antoine et Cléopâtre, et le troisième, 'Les Saturnales,' par celui de Tibulle et Délie. Le spectacle esquissé dans les papiers préliminaires a subi un processus d'apprivoisement; les images vigoureuses de jeux, d'assauts, de combats, ont cédé le terrain à une action nettement plus policée, imposée par les conventions chorégraphiques de l'époque. Le texte d'un spectacle grandiose imaginé pour fêter la majorité de Louis XV a été réécrit en divertissement d'été bien moins ambitieux.

Le compte-rendu du ballet publié dans *Le Mercure* de juillet en constitue une autre réécriture. Si Fuzelier ne le fit pas tout entier, tout au moins peut-on voir les traces de sa collaboration dans les trois pages du compte-rendu consacrées à la version 'royale' du texte.[21] Essentiellement un résumé du prologue et des trois actes joués à l'Opéra, cette analyse mentionne aussi en passant la mise en scène. A deux reprises, elle cite textuellement des phrases de l'avertissement du livret, soulignant dans la première l'originalité d'un sujet de ballet tiré de l'histoire.

En attirant l'attention du lecteur sur une version antérieure des *Fêtes grecques*, le texte du *Mercure* accentue l'instabilité des ouvrages de ce genre. En outre, il s'attache au travail d'élaboration par lequel le ballet dut passer avant de devenir texte parachevé. Cette façon de présenter le théâtre comme processus plutôt que comme produit fini se situait dans le contexte des premiers balbutiements d'une critique théâtrale qui n'existait pas encore en tant que discours autonome en 1723.[22] En l'absence d'un corpus de réflexions sur l'art proprement théâtral (c'est-à-dire, vu de l'angle scénique, et non pas littéraire), la rubrique des 'Spectacles,' créée dans *Le Mercure* de 1721 par Fuzelier et ses collègues, cherchait à produire l'effet de recul critique qui, selon eux, manquait au spectateurs de l'époque:

Il seroit a souhaitter qu'il parut souvent des Dissertations exactes, qui développassent l'art, les beautez, & les defauts d'une Piece, qu'un Spectateur ne peut pas toujours saisir dans la representation. ... Quoy que le Theâtre François ait de grands avantages sur ceux de nos Voisins, on ne sçauroit nier cependant qu'il n'y ait bien des defauts à corriger, ce qui ne se peut faire qu'en marquant les écueils de ce genre d'écrire.[23]

Marquer de tels écueils ne fut pas que le travail d'une critique théâtrale émergeant dans les journaux. C'était déjà l'un des objectifs principaux des prologues du temps de Fuzelier. Ces textes reflétant d'autres textes les réécrivaient en quelque sorte par anticipation, en proposant aux spectateurs un cadre enchâssant ou un registre détaché, atténuant ou évaluant le plaisir qu'ils allaient éprouver lors des spectacles ainsi présentés. Quant aux spectacles, perçus simultanément sur deux registres, ils devenaient l'expérience d'un flottement entre deux textes dont ni l'un ni l'autre ne pouvait être en fin de compte définitif. Si l'on en croit le prologue des *Fêtes grecques,* que doublait l'avertissement imprimé, et auquel faissaient écho les pages du compte-rendu consacrées à une autre version du ballet, le déploiement d'un tel appareil dialogique répondait à un besoin profond chez Fuzelier.

La parodie elle aussi entame un dialogue entre deux textes; dialogue que Fuzelier allait alimenter considérablement en écrivant *La Rencontre des opéra.*[24] Ce petit acte en treize scènes, constitue en lui-même un exemple frappant du phénomène de la mouvance textuelle. Cette soi-disant 'parodie' des *Fêtes grecques,* qui est bien davantage une réécriture de son prologue, se mue en dialogue sur un fait divers (la visite anticipée d'une troupe d'opéra italienne), fait divers qui s'élargit à son tour en appréciation de l'apport de la musique populaire, laquelle appréciation s'avère être le prétexte d'une scène de virtuosité vocale exécutée par un acteur habile.

Le décor de *La Rencontre des opéra* rappelle celui du prologue des *Fêtes grecques,* tout en le changeant de façon significative. Dans le ballet, la scène se caractérise par son aspect ordonné, monumental, et formel:

Le théâtre représente au fonds le Temple de Memoire orné des Statües des grands Hommes & d'Inscriptions à leur loüange.[25]

Lieu de la cérémonie publique, cet espace est investi par une troupe de danseurs rangés sous les ordres du dieu Apollon.[26] Par contre, l'espace de *La Rencontre* quitte la place publique des *Fêtes* pour l'intimité de l'école d'Erato. L'art (de l'Opéra) cesse d'être vu de l'extérieur pour l'être de l'intérieur; les trois actes des *Fêtes grecques* sont considérés du dehors comme unités détachées à suivre, le temple et les statues pour leur part symbolisent un 'déjà accompli,' alors que dans *La Rencontre* la création se voit du dedans comme répétition et composition, un processus en cours et non achevé.

Les différentes séquences de *la Rencontre* maintiennent le même effet d'irrésolution, soit en renversant ce qui avait déjà été admis dans *Les Fêtes grecques*, soit en multipliant des moments révélateurs de mises en abyme. La surprise causée par la présence de la Muse de l'histoire dans le prologue des *Fêtes grecques* fait de Clio une figure d'intruse. L'intrusion, réussie dans le ballet, échoue dans *La Rencontre* lorsque l'esclave d'Erato renvoie Clio sous prétexte que la fête des Saturnales dispense les musiciens de travailler. Le départ subit de l'Histoire prive la parodie des sujets du ballet tout en ouvrant la voie à une chasse aux textes pour combler la lacune. Cette chasse devient le prétexte d'un défilé de plusieurs personnages reflétant le théâtre (auteur, spectateur, etc.), chacun reprenant divers éléments des *Fêtes grecques*.

A titre d'exemple, le premier, un Député des beaux esprits, évoque le ballet sous l'angle de l'écriture de sa parodie. En dépit du scepticisme de Chriseïs, esclave d'Erato, le Député affirme qu'une parodie de ballet est concevable:

LE DEPUTE: Eh! bien pour abreger je crois que Malgré l'aridité de la Matiere la parodie d'un Ballet peut se risquer.

CHRISEIS: Et Moy je vous soustiens qu'il est impossible qu'elle soit Seulement Tolerable. Comment critiquer l'action d'une piece qui n'en a point? d'une piece ou chaque acte est composé de personnages nouveaux qui expedient deux petites scenes pour cedder le Terrain à la Danse? (sc. 4)

En même temps qu'il se livre à une réflexion sur son propre ballet, Fuzelier se livre à une réflexion sur cette réflexion. Deux scènes plus tard, Erato envisage l'écriture d'une deuxième parodie composée cette fois-ci, dit-on, par l'auteur lui-même des *Fêtes grecques:*

ERATO: [Chriseïs] pretend nous donner une Parodie composée par L'auteur même de l'ouvrage apostrophé

LE DEPUTE: Le Phenomene est rare!

ERATO: Et n'en sera peut-être pas plus brillant

LE DEPUTE: Eh! pourquoy Diantre cet auteur la a t il la demangeaison de se plaisanter luy même? tandis que tout Paris fourmille de poetes charitables qui sont tous prests a luy rendre ce petit service la avec plaisir. (sc.6)

Si *La Rencontre des opéra* ne vit jamais les feux de la rampe, *Les Saturnales* en furent illuminées les 2 et 3 septembre 1723.[27] Vite tombées, elles furent retirées de l'affiche des Comédiens Italiens pour réapparaître trois jours plus tard en un seul acte, sous le titre *Le Debris des Saturnales.*[28] Cette réécriture d'une réécriture transfère l'action dans une guinguette où défilent plusieurs personnages des *Fêtes grecques.* Sous le signe libérateur des Saturnales — et avec l'aide du vin de la guinguette — les actions et les personnages du ballet subissent des dévalorisations et/ou mises en évidence. On dit de Timée, 'Son esprit ne paroît guere ...', de la volte-face d'Antoine, 'Vous aves eté l'amour de Cleopatre aussi promptement que son Juge,' et de Tibulle, 'il a l'esprit un peu trop lourd.' Ces critiques s'accompagnent de clins d'oeil au public, comme au moment où Cléopâtre repousse Marc-Antoine en disant: 'on nous regarde au moins' (sc. 9).

Le groupe de textes relevant des *Fêtes grecques et romaines* a illustré une pratique de la mouvance, trait on ne peut plus théâtral. D'autre part, les différents documents que nous avons considérés ont livré plusieurs cas d'une tripartition dynamique de l'événement théâtral en texte achevé, en foyer d'analyse critique, et en procès créateur saisi sur le vif. Au niveau de la réception, ce morcellement se traduit en impressions de flottement plus ou moins continu: flottement de motifs à grouper; flottement des conventions de l'opéra; flottement des genres 'opéra' et 'ballet;' flottement du discours critique entre la préface, le prologue, le compte-rendu, et la parodie. ...

Depuis l'ébauche 'Pour le Roy ...' jusqu'au *Debris des Saturnales,* il a été possible de suivre des transformations successives des *Fêtes grecques.* L'on peut dire que ces transformations correspondent aux étapes chronologiques impliquées par les termes *avant-texte, texte,* et *après-texte.* Grâce à un processus de parachèvement de l'écriture, les brouillons de Fuzelier atteignent une clôture précaire de texte au moment de la présentation de son ballet. L'événement que fut la mise-en-scène des *Fêtes grecques,* le 13 juillet 1723, au théâtre de l'Opéra, déclencha aussitôt une nouvelle série de réécritures: le texte du compte-rendu du *Mercure,* le manuscrit de *La Rencontre des opéra,* la parodie des *Saturnales,* et la compression rapide de cette dernière en un *Debris des Saturnales.* Autant de documents textuels — existants ou ayant existé — qui marquent les moments d'arrêt d'un mouvement diachronique vers l'avenir, mouvement qui reflète en miniature l'évolution du théâtre à cette époque.

Mais l'intérêt du cas des *Fêtes grecques* ne se réduit pas seulement à l'alignement dans le temps d'une série de documents écrits. Presque tous ces textes renferment les éléments d'un discours réflexif et/ou autocritique, précondition nécéssaire de toute évolution, et qui commence par la feuille 'Pour le Roy ...' dont les ratures constitueraient déjà, d'après Bellemin-Noël, un geste d'écriture critique: 'Toute rature met en cause l'ensemble de l'écriture.'[29] L'avertissement et le prologue des *Fêtes grecques* se combinent ultérieurement pour renforcer ce registre critique du ballet: tous les deux soulevant la question d'une thématique convenable à l'opéra, ainsi que celle du rôle de la danse dans les productions d'une Académie de Musique. La fonction critique naissante des articles publiés dans *Le Mercure* n'a pas besoin d'être tirée en longueur ici. Pour ce qui est de la fonction critique des parodies, il est significatif que *La Rencontre des opéra* l'ait accentuée en choisissant le prologue des *Fêtes grecques* pour son évaluation du ballet.

L'article du *Mercure* envisageant l'établissement d'une critique théâtrale soulignait l'importance d'un recul pour saisir les 'défauts d'une Piece, qu'un Spectateur ne peut pas toujours saisir dans la représentation.'[30] Il laissait supposer un processus à deux temps; celui d'une réception des impressions immédiates de la représentation, et celui, ultérieur, d'une 'relecture' critique. Dans une certaine mesure, les *après-textes* de Fuzelier (tant journalistiques que dramatiques) permettaient une telle relecture. Toutefois, tous les textes considérés ici offrent à leurs lecteurs/spectateurs la possibilité de dépasser une reception trop linéaire, et de faire coïncider les deux opérations. La présence d'un certain nombre d'éléments auto-référentiels déclenche un jeu réciproque de perceptions qui oscillent entre la réception du texte comme produit fini et la réalisation qu'il n'est qu'un procès en cours.

On a pu assigner aux brouillons un statut d'*avant-textes* grâce à leur clôture relative. Par contre, les renvois à des 'non-réalisés' tels que l'énoncé 'voir au sujet des festes des anciens,'[31] ou 'on peut mettre en action ...' leur donnent un caractère inachevé et ouvert. De même, la dernière phrase, au futur, de l'avertissement des *Fêtes grecques*, 'on n'apprendra que trop tôt si on s'est égaré,'[32] ainsi que l'assemblage par Apollon dans le prologue des ingrédients (chant, histoire, et danse) du spectacle à suivre, ont pour effet de valoriser le dynamisme et non pas la fixité de l'événement.

En fait, la mise en forme que pouvait constituer l'impression / représentation des *Fêtes grecques* n'était qu'une mise en forme relative. Sans parler des différences fondamentales entre le 'texte' que fut le livret imprimé, et celui que fut sa première réalisation scénique, les citations dans *Le Mercure* venant d'une version antérieure du ballet, la réécriture des paroles de la chanson qui accompagna le début de Mlle Richalet, et surtout, l'addition d'un nouvel acte, 'La Fête de Diane,' à une reprise des *Fêtes grecques* en 1734, font que le 'texte' du ballet héroïque ne se fixe pas de façon définitive. Il n'est pas étonnant alors que la phase appelée *après-texte* ne suive pas tant *Les Fêtes grecques* qu'elle l'accompagne dans le temps, au point d'être inextricablement mêlee au processus originel d'écriture. L'avertissement qui précède le livret et le prologue qui annonce le ballet propre ont entamé le discours autoréférentiel avant même que le texte ne soit reçu par le lecteur/spectateur et, par là, rejoignent au-delà du texte les procès d'écriture en cours que renferment le compte-rendu et les parodies.

L'extraordinaire application que Fuzelier mit à réécrire son texte a de quoi étonner. Cependant, notre propos n'est pas de chercher à savoir s'il s'agit d'opportunisme ou de quelque autre projet plus noble. ... Ce qui importe c'est que la série de textes associés au *Fêtes grecques* offre une illustration segmentée du geste complexe de la production/réception du fait théâtral. Pendant une période de quelques mois à peine, Fuzelier aborda son sujet plusieurs fois, chaque fois d'un point de vue différent. Il fut auteur, critique, spectateur de lui-même. En cela il fut représentatif de son temps, car le dix-huitème siècle français vit s'épanouir une véritable manie du théâtre: les sujets passaient d'une forme à une autre, les spectateurs d'un théâtre à un autre, les practiciens d'un registre à un autre, l'art du spectacle — en quittant la scène pour faire irruption dans la vie — passait d'un monde à un autre. La poussée de l'imprimé, à laquelle contribuait Voltaire, et qui relevait par sa tendance vers l'édition 'définitive' d'un certain goût classique de la transparence et de la permanence dans l'expression, fut contrebalancée par la mise au point d'un mode de communication plus ambigu, fait des nuances, des non-dits, et des qualifications d'un langage en évolution constante. L'ampleur attestée par les textes de Fuzelier d'un dynamisme ennemi de formes contraignantes entraîna à son tour l'exigence d'une nouvelle forme capable de fixer, sans pour autant l'immobiliser, ce caractère de mouvance, c'est-à-dire, capable d'être objet (texte) et procès (écriture/réécriture) simultanément.

Si la postérité nous apprend que Louis Fuzelier n'y a pas entièrement réussi, l'exemple du cas des *Fêtes grecques et romaines* reste néanmoins éclairant pour qui voudrait comprendre le fontionnement de la mouvance, ou de la théâtralité, dans tel texte du théâtre de Marivaux. Accusé de n'avoir fait que des réécritures sans nombre du même sujet, Marivaux semble effectivement s'être répété en puisant régulièrement dans le fonds conventionnel du théâtre italien et en employant deux fois le titre *Surprise de l'amour.* Cependant, son œuvre est un entrelacement subtil de textes conventionnels, de création originale et d'interrogations sur elle-même en tant que théâtre. Là où il faut faire face dans une pièce telle que *Les Acteurs de bonne foi* à une concentration de tous les types de passage (entre deux 'textes') dans un seul acte, la fragmentation du phénomène intertextuel en huit textes séparés chez Louis Fuzelier facilite l'analyse d'une mouvance autrement insaisissable.

DAVID TROTT
Erindale College

Notes

1 *Le Mercure*, juillet 1723, p. 134.

2 'Voici une Parodie faite pour elle sur l'air qu'elle danse, habillée en Flore à la suite de Mlle Prevost' (*Le Mercure*, septembre 1723, pp. 581-2). Les paroles de la 'parodie' évoquent une danse dans laquelle l'élève se serait efforcée d'imiter les pas de son mentor. Il faut se rappeler qu'en 1723 le terme 'parodie' gardait encore son sens de 'paroles réécrites sur un air connu.'

3 *Catalogue des imprimés de la Bibliothèque Nationale*, 'Voltaire,' 2 volumes (214-I et 214-II), 1978, p. xii. Pomeau dit que Voltaire est 'l'auteur ... le plus copieusement représenté au Départment des Imprimés' (p. xi). En effet 'Voltaire a vécu pour écrire, et pour imprimer' (p. xii). Lorsqu'en 1725 il apprend la parution de trois éditions pirates de son *Hérode et Mariamne*, il en parle dans les termes suivants: 'Je ne sais à cela que de donner ma piece et d'y corriger le plus de choses que je pourai, afin que l'air de la nouveauté soit joint à la correction dont elle avait besoin' (Best. D 243, 23 juillet 1725). La répétition de 'corriger' et 'correction' suggère qu'un texte, à l'étape de sa représentation, est imparfait. Cette imperfection peut être attribuée, soit aux erreurs de celui qui copiait les répliques *à vive voix*, soit à la nature même de la représentation qui, pour réussir, exploite une 'nouveauté' que l'édition ultérieure aura soin d'éliminer ou d'atténuer.

4 Paris: E. Ganeau, 1721-1737, 10 vol.

5 Pour une discussion plus détaillée du très grand corpus des écrits inédits de Fuzelier, voir mon article 'Louis Fuzelier et le théâtre: vers un état présent,' *Revue d'Histoire Littéraire de la France*, vol. 83, no. 4 (juillet/août 1983), 604-17.

6 *Le Texte et l'avant-texte*, Collection 'L', Paris: Larousse, 1972, p. 132.

7 Diderot voyait bien le pouvoir communicatif du théâtre. Il s'agit du passage célèbre où Dorval, conscient de 'l'utilité des théâtres,' rêve de les transformer en églises, et les acteurs, en prédicateurs: 'Ah! mes amis, si nous allons jamais à la Lampedouse fonder ... un petit peuple d'heureux, ce seront là nos prédicateurs; et nous les choisirons, sans doute, selon l'importance de leur ministère. Tous les peuples ont leurs sabbats, et nous aurons les nôtres. Dans ces jours solennels, on représentera une belle tragédie...' (*Entretiens sur le Fils naturel*, second entretien, Paris: Garnier-Flammarion, 1967, p. 52).

8 ms. Archives Nationales, AJ^{13} 1035(4), dans une chemise intitulée 'Extraits, fragments ... pouvant servir au plan d'opéras, de ballets... .'

9 Parmi les sujets abandonnés, se trouvent: 'les festes de julie,' 'une feste françoise,' 'une feste ... persane.' Utilisés ailleurs, furent 'les ages' (représentés en 1718), 'la foire des fées' (en 1722), et 'la boitte [sic] de pandore' (en 1721).

10 Répété treize fois.

11 'Je [Terpsicore, Muse de la danse] scauray mieux que vous amuser
    sur la scène
Le plus charman des Rois
...
Il Regarde vos jeux mais il se mesle aux miens.'

12 Des remarques servant d'aide-mémoire conseillent de 'voir l'explication des fables,' ou bien de 'voir au sujet des festes des anciens.'

13 Ayant effectué les recherches envisagées dans la note 12, Fuzelier donne quelques détails sur divers jeux romains et grecs: 'jeux Capitolins,' 'jeux pirriques,' 'jeux istmiens,' etc.

14 Quatre actes sont envisagés: I. 'feste grecque/ les jeux olimpiques ...'; II. 'feste Romaine/ les baccanales ...'; III. 'feste grecque/ les jeux istmiens ...'; IV. 'feste Romaine/ les Saturnales ... .' Le troisième acte sera supprimé par la suite, et le deuxième, combiné avec 'les festes égiptiennes/ antoine et Cleopatre,' ce dernier motif figurant ailleurs sur la feuille.

15 'Ce ballet avoit eté composé pour être representé sur le theatre des Thuilleries. L'auteur animé de l'honneur D'amuser le Roy dans un temps ou il vouloit bien embellir les spectacles, en Daignant s'y meler luy même, avoit imaginé de l'amener dans un Divertissement digne d'un aussi grand roy ...' (Duchaiseau, 'Histoire des théâtres,' ms inédit en 6 cartons, article sur Fuzelier, p. 6. Voir Bibliothèque de l'Opéra, Rés. 925).

16 A Paris, Chez La Veuve de Pierre Ribou, MDCCXIII.

17 '*Les Fêtes Grecques & Romaines* forment un Ballet d'une espece toute nouvelle ...' ou bien, 'On s'est écarté d'une route frayée depuis longtemps ...' ('Avertissement,' pp. iii et vi).

18 A la différence de la plupart des auteurs de ballets qui cherchent l'inspiration 'dans les Champs trop souvent moissonnez de la Mythologie & du Roman,' Fuzelier se tourne vers l'histoire, plus 'digne d'occuper les génies amateurs du vraisemblable' (*ibid.* p. iii).

19 La Vérité n'est pas toujours si redoubtable;
    L'Histoire aussi bien que la Fable
Peut fournir à vos chants des Héros amoureux.
(prologue)

20 APOLLON. – C'est en vain qu'aujourd'hui des chants mélodieux
    Sur la Scene appellent les Graces:
Si la Danse n'amuse & ne charme les yeux
L'Ennui suit les Plaisirs & vole sur leurs traces.
(prologue)

21 Fuzelier a vraisemblablement écrit tout le compte-rendu. Alors co-rédacteur du

Mercure avec Dufresney et La Rocque, il a dû répondre lui-même à cet appel que les éditeurs avaient lancé aux auteurs en 1721: 'Comme personne n'est plus capable, generalement parlant, de faire l'Extrait d'une Piéce de Theâtre que celui qui l'a composée, nous avertissons aussi les Auteurs qui doivent craindre qu'un autre ne les fasse pas parler & penser aussi bien qu'ils croyent avoir fait, de composer eux mêmes les Extraits de leurs Pieces, & de nous les faire venir ensuite, leur promettant de les insérer dans notre Journal tels que nous recevrons ...' (*Mercure,* novembre 1721, p. 4).

22 Les *Lettres historiques à Mr. D*** sur la nouvelle comédie italienne* (1717-19) de Boindin et les *Réflexions critiques* (1719) de l'abbé Dubos constituent deux exceptions importantes.

23 *Le Mercure,* juin et juillet 1721, pp. 131-2.

24 Ms. B.N. f.fr. 9333.

25 Prologue des *Fêtes grecques et romaines.*

26 'Les Eleves d'Erato s'y trouvent rassemblez par l'ordre d'Apollon pour seconder les desseins de la Muse de l'Histoire' (prologue).

27 Texte disparu. Le manuscrit portant le titre 'Les Saturnales' (ms. B.N. f.fr. 9332) est en réalité 'Le Debris des Saturnales.'

28 Voir *Le Debris des Saturnales.* ms. B.N. f.fr. 9332. Ce texte se trouve également dans *Le Dictionnaire des théâtres de Paris,* vol. V, pp. 39-55.

29 *Le Texte et l'avant-texte,* p. 5.

30 Voir page 81, note 23.

31 Voir note 12.

32 Avertissement des *Fêtes grecques,* p. vi.

6. Mme Bégon, a Colonial Salon Hostess

To students of literature and eighteenth century social history the word 'salon' is evocative of those brilliant evenings at Sceaux or in Paris, frequented by the philosophers, artists and writers of a period when France was the centre of the civilised world. Writing of Mme de Lambert, the first of the eighteenth century hostesses in 'ce siècle où les femmes régnèrent par les salons,' Mathurin de Lescure says, 'elle fut une puissance sociale, littéraire, académique, dirigea la mode, régenta le goùt, imposa le ton, fit de son éventail le sceptre de la conversation, donna de ces diners dont le billet d'invitation était le brevet de réputation et d'influence.'[1] Elisabeth Badinter, writing of Mme de Châtelet, says 'Elle aurait aimé réunir autour d'elle et Voltaire, à Cirey, une sorte de petite Académie des sciences. Connaissant l'élite de la pensée scientifique européenne, il n'aurait pas été impossible de réaliser ce rêve.'[2]

Such were the hostesses and salons that achieved fame, but there were others of lesser stature, which yet conform to the Littré definition of a salon as ' ... La maison où l'on reçoit habituellement compagnie, et particulièrement, bonne compagnie, et où l'on cause. Par extension: la bonne compagnie, les gens du monde.'[3] It might be assumed that a colonial salon, because of its isolation from Paris, would fall into this more general category. But New France had its miniature capitals where women undoubtedly 'reigned.' Of such hostesses, Mme Bégon was a notable example.

Elisabeth Bégon was born in Montreal in 1690 and died in 1755; she was an almost exact contemporary of the two Parisian hostesses, Mme du Deffand and Mme Geoffrin, whose lives most resemble hers. Like them, she wrote letters[4] which reveal both her personal drama and much

MAN AND NATURE / L'HOMME ET LA NATURE III/1984
0824-3298/84/0300-0089 $01.25/©C.S.E.C.S./S.C.E.D.S.

of the social life of her day. Like theirs, her correspondence is of interest as a portrait of the personalities, manners and values of the age. Yet the geographical and historical characteristics of her milieu are very different from theirs. So, if comparison between the salons of Paris and the salon of Mme Bégon is to be made, certain characteristics of the colonial setting must be sketched.

The first significant difference between the milieu of Paris and that of Mme Bégon is scale. Mme Bégon's world was at once far bigger and far smaller than that of the capital. In terms of population, the urban centres of Montreal, Trois-Rivières and Quebec were very small indeed, but they provided the political and social direction and necessities of life for an immense geographical area, covering much of North America and the West Indies. Whereas the habitués of Paris salons looked intellectually afar and corresponded with England, Germany and Russia, but found Mme Geoffrin's journey to Poland 'une fantaisie des plus excentriques,'[5] Mme Bégon's acquaintances travelled incessantly. The smallness of her immediate circle and the vastness of the distances surrounding it give to her evocation of New France a curiously kaleidoscopic effect: distance is constantly diminished by people transcending it and bringing news. News from next door and news from abroad, from the distant forts and adjoining colonies, news of friends and news of the enemy form the stuff of Mme Bégon's letters, accounting partially for the fact that her letters, never intended for publication, have become primary sources of Canadian history.

Mme Bégon was writing at a critical period. 'La Nouvelle France est à son déclin. Au milieu des plaisirs des grands et de la misère des petits, elle s'achemine progressivement vers la débâcle de la guerre de Sept Ans.'[6] Threatened by war, by famine, by cold and by want, the life-style of New France was one where insecurity, personal and collective, was the human condition. Mme Bégon's life parallelled the hopes and despairs of her time. She was young during the last phase of growth and hope in the thirty years' peace. In her maturity, she knew the corruption, the scandals and the decline of New France. In her old age, she was to see the first stages of conquest, with the loss of Acadia, the twice repeated downfall of Louisbourg. Although she did not live to see the fall of New France she foresaw disaster, her insights being keen and politically acute.

Hers was, indeed, a political milieu. She was born and married into the governing classes of New France. Etienne de Rocbert de la Morandière, her father, was 'gardien du magasin du roy.' Her husband, Claude Michel Bégon, the younger son of a noble family from Blois, came from a line of administrators of New France. He became a Chevalier de St. Louis, thus acquiring the most coveted rank of his world, although his highest post was as Governor of Trois Rivières. Her brother-in-law, Michel Bégon, became Intendant of Canada and married the sister of the Minister of Marine, the Conte de Maurepas, second only to the king in authority over New France. Her father-in-law, was Governor of the French West Indies and Intendant de la Généralité de La Rochelle. Her son-in-law, Honoré Michel de Villebois de la Rouvillière, to whom the letters are addressed, was, at the time, in Louisiana, newly promoted from Commissaire Ordonnateur de Montréal, whereas the most powerful man in her circle, the Marquis de la Galissonière, who was her nephew by marriage, was acting Governor of the colony in the absence of the captured La Jonquière.

Not all these connections were helpful to her, however. The Bégons and their powerful relative, Maurepas, considered that Claude Michel Bégon had made a *mésalliance,* and their marriage occasioned a power struggle between the Governor and the Bishop of Quebec. Such a contretemps was doubtless part of the political education of Elisabeth Bégon who came to pride herself on her discretion and tact, and apparently with reason, since there came to her house everyone of note in New France, including those whom historians admire (La Vérendrye, Hocquart, La Galissonière), those who provoke controversy (Rigauld de Vaudreuil, for example), and those whom history has judged severely, partly because of her judgment (Varin the corrupt, Longueuil the pompous, and Bigot the extravagant).

These are the 'gens du grand monde,' 'les mondains' and 'les personnalités diverses,' the guests of her salon. They are, in the Canadian sense, aristocrats, defined by Chanoine Groulx as 'tout simplement l'homme de classe sociale, noble, seigneur, officier de troupe, de justice, fonctionnaire du gouvernement ou de l'administration, qui se situe au-dessus de la bourgeoisie.'[7] Not thinkers or artists, though some were amateur scientists and well educated, they were preoccupied with practicalities. Their task was to preserve the peace with the Indian allies, to protect trade for France and to hold the colony as a political and religious entity against the growing power of the English, their ambition to make a fortune or at least a solid career with a pension. Few had the vision to transcend the immediate situation and they were not encouraged to do so. Orders and directives came by every mail from Versailles, making it clear that every aspect of life was controlled by the king. Yet there were long periods each winter without mail or news, periods when the land and the climate marked the inhabitants as left to their own resources. This isolation accentuated their strengths and weaknesses, focusing a merciless light on them individually and collectively.

Mme Bégon observes her guests lucidly and, like her Parisian contemporaries, adds her personal touch to her portraits. Since her intention is to amuse her son-in-law, she indulges a taste for light satire, and, because her vision is family-oriented, she presents her acquaintance on a domestic scale. Historical figures appear in unexpected postures as a result. La Vérendrye, home from establishing forts all across the West, is caught flirting in the salons of Montreal where 'C'est la plus jolie chose du monde de voir la mine que La Vérendrye fait à la Martinière.'[8] The worldly Bigot is depicted standing with his hands behind his back, apparently bored, while all Montreal dances in his honour. 'S'il danse deux ou trois menuets, c'est tout.'[9] The Baron de Longueuil, whose promotion had been delayed during the life-time of Mme Bégon's husband, is shown 'dans un état triste ... par l'incertitude où il est de savoir s'il sera gouverneur ou non.'[10] Another day, the Baron arrives drunk at Mme Bégon's house and invites himself to dinner, while M. Varin, the Commissaire Ordonnateur de Montreal, who had succeeded 'le cher fils' in that post, falls down, drunk also, 'sa perruque d'un côté, et lui de l'autre.'[11] Priests, bishops, judges, governors and their ladies, are all evoked by her satirical pen, along with the marriages, deaths, gossip and political news. She observes of it all: 'C'est une vraie comédie que le monde.'[12]

This society was small. Guy Frégault estimates their number at two hundred and eight,[13] while Isabel Landels calculates that Mme Bégon mentions three hundred names.[14] These were the upper classes of urban New France, and of its three towns: Quebec, Trois-Rivières, and Montreal, in all three of which Mme Bégon lived at one time or another. Miniature capitals – ports, garrison towns, centres of law and trade, religion and pleasure – they were, for travellers, a refuge from the hardships of the wilderness, and for their own population a dynamic network of social, political and commercial life. Above all, the towns of New France were parishes and groups of homes. The home, the multipurpose building, was the setting for most human activity, including off-duty military life, trade and government. It was in the homes that the majority of public business was conducted and there too that neighbours, colleagues, and visitors were received, fed and entertained. The living rooms, the reception rooms – the salons – were thus at the centre of colonial life.

The domestic setting apparently influenced the style of the receptions. Unlike those of Paris, where 'on fondait un salon au lieu d'aller à l'eglise, on causait au lieu de prier,'[15] they were rarely deliberately planned. One has the impression from Mme Bégon's letters that the receptions were often spontaneous and improvised according to circumstance: to celebrate the arrival of a traveller, to honour an official, or to entertain friends and colleagues after the working day. Louis Franquet has left a description of his welcome by the Governor of Trois-Rivières who succeeded M. Bégon in that post: 'M. le Gouverneur voulut absolument me conduire chez lui: il fallait cèder à ses instances. Y arrivé, je fus présenté à madame son épouse, qui par parenthèse, est une personne des plus accomplies tant par la figure que par l'esprit. Elle est d'ailleurs pleine de grâces et de politesse; ... L'on ne tarda pas ensuite à passer dans la salle à manger. Il y avait une table de vingt couverts servie, je ne dirais pas comme à Paris, d'autant que c'est l'endroit où j'ai vécu le plus frugalement, mais bien avec la profusion et la delicatesse des mets des meilleures provinces de France. On y but toutes sortes de vin, toujours à la glace; jugez du plaisir par le chaud excessif qu'il faisait. Après le dîner, fait une partie de cadrille.'[16]

Franquet's description reveals the sort of entertainment that was offered spontaneously in a home in New France: a good dinner, plenty of wine and a hand at cards to follow. Testimony abounds, and particularly in Mme Bégon's letters, that dancing and gambling were also popular. Claude de Bonnault writes of a prevailing feverish *joie de vivre* and desire for amusement: 'Comme si l'on n'était pas sûr du lendemain, parce que le risque pour ces hommes était pain quotidien, on se pressait de jouir de tout. On voulait vivre de toutes ses forces, on lisait, on dansait, on jouait aux cartes avec frénésie et l'on aimait. On a beaucoup aimé au Canada en ce temps-là.'[17]

Elegant conversation and discussion of the arts also formed part of the entertainment in the Canadian household. Charlevoix found in New France 'un petit monde choisi où il ne manque rien de ce qui peut former une société agréable,' where 'on politique sur le passé, on conjecture sur l'avenir, les sciences et les beaux arts ont leur tour, et la conversation ne tombe point.'[18] Pierre Kalm noted that 'les gens de distinction en général ont bien plus de goût pour l'histoire naturelle et les lettres que dans les colonies anglaises.'[19] Frégault states that in the seventeenth century the plays of Corneille and Racine had been performed in New France 'peu après avoir été jouées en France, et on ne manque pas de lectures, ainsi que le démontrent les inventaires de l'époque.'[20] Mme Bégon confirms the taste for reading that must have formed the subject of some discussion. Her grand-daughter, she writes to 'le cher fils,' reads all day long, and she lists the books she has bought: 'tous les Corneille de l'univers, les Fables, les Henriade, les Don Quichotte, les Ducerceau' and 'mille dictionnaires latin et français.'[21]

A certain level of culture thus existed in the society of New France, but estimations of its degree of refinement vary. Contemporary evaluations were on the whole favourable. 'Les habitants sont affables et de bonne société,' wrote Jean Bonnafoux,[22] and Kalm found the Canadian politeness 'plus raffinée que celle des Hollandais et des Anglais des colonies appartenant à la Grande Bretagne.'[23] Charlevoix found 'des cercles aussi brillants qu'ailleurs,'[24] but Céline Dupré, commenting upon Mme Bégon's letters, remarks that 'ce ne sont pas toujours les plaisirs les plus raffinés que l'on recherche avec tant d'ardeur, en société. A vrai dire, l'aristocratie "montréaliste" semble plutôt rude de moeurs, du moins à en juger par sa façon de festoyer. Ses dîners se transforment souvent en d'interminables gueuletons où l'on "chante sauvage" et boit au point d'avoir peine à danser un menuet.'[25]

Refined or less so, and certainly without the distinguished wit of a Voltaire or a Marivaux or – the ultimate accolade of the Parisian salon – an academician, the salons of New France were probably, above all, occasions for the exchange of news and conversation about practical matters. In Mme Bégon's letters, military matters, politics, problems of public morality jostle with intimate jokes, descriptions of the weather and family gossip, leading one to imagine the same mixture of topics in the salons. In these, the strength of outlying forts would have alternated with the ethics of profiteering in flour or the inevitability of liquor smuggling, the illegal trade in furs with the potential of coal mining, the state of the cod fishery with the possibility of growing flax and teaching the 'habitants' to spin. The news of marriages, illnesses and deaths would have been mingled with the need to build mills, to establish a ship building industry, and to replace furs as the staple product of the colony. In the spring all the talk would have been of promotions and the resultant shifts in the rival cliques and power alliances. Then there would be debates about the struggles between church and state, about the right of soldiers to marry, and about the political or commercial role of missionaries in distant forts. There was tension in the military hierarchy and jealousy between the troops of the various settlements, the French despising the colonials and the colonials despising each other in proportion to their distance from authority. The clergy too indulged in endless 'chicanes de précédance' and, if other quarrels were in abeyance, there was the complex rivalry among the Indians to consider and the ever present tension between the governor and his intendant. Each of these topics would have had its partisans, and the arguments about policy and development would have been sprinkled with discussion of the latest measure to prevent fire, the by-laws concerning the construction of houses, or the properties of medicinal herbs. Always the talk would have veered to the shortage of money.

Shortages – of men, of arms, of food, of pay – famine and want, these were the realities behind the brave show and the feverish *joie de vivre.* Mme Bégon's letters reveal poverty at all social levels. She remarks of two households she visits: 'Il y a un air de disette dans ces deux maisons pitoyable.'[26] As a widow without a pension, she has difficulty in making ends meet and much of her correspondence concerns her financial difficulties and comparisons with her former state. Yet Franquet, speaking of the incumbent Governor of Trois-Rivières, says that he had an income of '1000 livres au plus. Cela ne suffit point, d'autant que cette ville étant située à moitié chemin de Québec à Montréal, sa maison est le rendez-vous de tous les passagers, et il est si honorable et si généreux qu'il y reçoit le petit comme le grand.'[27]

The shortages of food and money, though not unknown to some of the salon hostesses of Paris, were undoubtedly a significant factor in the life-style of those of New France, and a determining influence on their role. Most ladies must have known the reality of Mme Du Deffand's 'recette contre l'ennui ... l'exercice du corps, l'application de l'esprit ou l'occupation du cœur.'[28] Most had to do some of the house-work. Mme Bégon refers to herself knitting, cooking, gardening,[29] doing her own decorating,[30] supervising the workmen building a new wing to her house,[31] renting it to Bigot[32] and – despite her diffidence as a business woman – learning the financial management of a vineyard in France.[33] In this housewifery, she was not exceptional. At one point she mentions visiting the wives of an ex-Governor and a military commander and finding one of them darning her husband's clothes.[34]

An ability to conjure hospitality and entertainment from an inadequate income obviously belonged to the role of the hostesses in New France. Yet this did not lessen their influence as leaders in the art of gracious living. Franquet's description of his hostess as an accomplished lady is echoed by other visitors. Kalm says 'La Québecoise est une vraie dame française par l'éducation et les manières.'[35] He remarks that 'les belles dames du Canada ne peuvent entendre aucun barbarisme ni expression inusitée sans rire.'[36] Charlevoix was gratified by the purity of the language spoken in New France: 'Nulle part ailleurs on ne parle plus purement notre langue.'[37] Franquet endorsed this: 'ils parlent un français épuré, n'ont pas le moindre accent,'[38] and Montcalm observed that 'les paysans canadiens parlent très bien le français.'[39] Bougainville mentioned that 'ils parlent avec aisance ... leur accent est aussi bon qu'à Paris.'[40] That this correctness of speech was the work of the ladies, who taught it along with social conduct, is confirmed by Mme Bégon whose granddaughter recites to M. de La Galissonière 'presque tous les jours quelques fables ou petits vers.'[41] Superintending the children's education was an important task of the lady of the house. Latin, reading, correct behaviour, religious reverence, social tact and poise without haughtiness are the aspects Mme Bégon reports most frequently to the absent father.

Mme Bégon herself, indeed, defines her role in life as that of a mother. It is as such that she signs all her letters, yet, like Mme Geoffrin, she has 'l'humeur donnante'[42] and so includes in her concerns every one who asks for her help. She puts at their disposal her good connections, her personal qualities of kindness, discretion and common sense, exercising in doing so considerable influence. As she reminds her son-in-law, 'L'estime que ta mère s'est acquise dans le pays lui donne la confiance et l'amitié de personnes qui peuvent beaucoup.'[43] Visited almost daily by the Governor, she is courted by friends and enemies alike in their search for a pension, a promotion, the ear that could lead to favour in Versailles. Receiving courtesy calls, answering requests, arranging introductions, writing references, despatching clothing, comfort and advice are what she calls her 'métier.' On May 1750, she wrote sixty letters before sending a note to her son-in-law. Though often wearied by it all, she can still smile ruefully at the end of a long day: 'J'ai fait des visites tout le jour, dont je suis si lasse que je me donnerais pour deux liards. Quel métier, cher fils, et surtout quand il faut essuyer les compliments de deux communautés.'[44]

It is in her letters concerning her immediate family, however, that we see most clearly her influence on manners and the values that she practised. Her fundamental attitude is one of moderation: in matters of religion, education, work and leisure, she rejects extremes and absolutes, revealing a tendency to be governed by pragmatic courtesy and common sense. She is, for example, devout and regular in church attendance, because she derives consolation from her religion. But she will laugh with her grand-daughter at a dull sermon, and is amused by the hypocritical 'grosses-têtes' bowing to the clergy's thunderous threats of excommunication while all Montreal is dancing in honour of Bigot's visit.[45] At home she encourages her grand-daughter to read, despite critics who consider that the little girl should learn house-work: 'Je la laisse faire, persuadée que cela lui cultive l'esprit.'[46] Yet in the matter of her grand-daughter's social manners, she is more adamant. She will not tolerate 'la hauteur' either in her charges or in others. Like Mme du Deffand, she could claim 'ma folie n'est pas la présomption ni la prétention.'[47] Like Mme Geoffrin, 'elle dédaignait toute espèce de pose et d'affectation.'[48] It is one of the comparisons between Canada and provincial France that she makes most often, deploring the French preoccupation with rank and grandeur. She is particularly hurt by the arrogance of her sister's family: 'ce sont des airs de grandeur et de hauteur avec nous qui me font pitié,'[49] and 'ils sont tous d'une gloire insupportable.'[50] In a sad letter, she writes of France: 'ce pays n'a d'aimable que le climat. Du reste, il est pire que le Canada pour la gloire, l'envie, la jalousie et tout ce qu'il y a de moins bon dans la société.'[51]

Tactlessness and lack of common sense are what she deplores most in her son-in-law when he develops a persecution mania which she does not understand. She scolds him severely for his reckless and haughty behaviour, then, to cure his 'mille chimères dans la tête,'[52] she offers him advice for success in his career. While acknowledging his scrupulous honesty, she considers it too absolute and urges him to remember his dependents and to make some money. 'Si tu n'as pas l'esprit de gagner quelque chose où tu es, tu mériterais d'être battu, car on ne se fait plus de mystère aujourd'hui de ce métier, et on regarde comme bêtes ceux qui n'ont pas de commerce en tête. Je crains que cela ne t'occupe pas assez.' Lest he find this too severe, she adds, 'Il est vrai qu'il convient en quelque façon beaucoup mieux de s'occuper de son devoir, mais il faut faire en sorte de travailler à l'un et à l'autre.'[53] She reveals a similar sense of bleak reality when she summarises the situation of the colonial civil servant: 'Il faut attendre une place ou être en état de vivre sans le secours du roi.'[54]

These qualities of realism and tact together with her social connections and her wit and charm made of Mme Bégon a hostess in the Mme de Lambert, Mme Geoffrin manner. Hospitality was obviously a way of life in New France, and we can deduce that, during her husband's life-time, Mme Bégon's receptions were stylish, lavish and entertaining, with the arts of *savoir vivre* in the eighteenth century grand manner practised in them. Yet doubt remains as to whether she can be said to have had a salon as literary historians use the word, for the single most significant element of the Parisian salon world was lacking: the deliberate, self-conscious cult of fame and intellectual eminence. Mme Bégon did not entertain to combat ennui nor to ensure personal recognition, nor did she seek to dazzle by her guest list or her theatrical recreations. She did not aspire to celebrity as a *femme savante* or a friend of the great. Like many another hostess presiding over 'la bonne compagnie, les gens du monde,' she was practising the rites of courtesy and fulfilling the obligations of hospitality at a time and in a society where these were an intrinsic part of civilisation.

That she did so with wit and warmth and that she was surrounded by the men who made the history of eighteenth century New France ensured her a posterity she did not seek and a readership she never intended. Her destiny was to leave a record of her society so clearly marked with the imprint of her personality as to have influenced history's vision of her world. 'Sur cette période troublée de notre histoire,' writes one of her biographers, 'la correspondance de Mme Bégon ne jette pas tellement une lumière nouvelle, susceptible de révéler des faits inconnus ... Ce qu'elle offre de précieux, c'est la découverte du contexte humain.'[55]

The human context of her portrait of New France gives it a luminosity quite different from the lustre and artifice that literary historians have attributed to the salons of Paris. It is that of a small society in a vast land, condemned to defeat, but engaged in a valiant task – that of defending an empire and supporting social values. Twin aspects of a single endeavour, both were presided over by a woman whose role as a mother was part of a larger social avocation, that of making life around her as gracious, sensible and kindly as possible.

For that finally was the achievement of the hostesses of New France, as it was of the less intellectual hostesses of Paris – to soften, by common sense and friendship, the harshness of destiny. 'Ne dois-je pas,' asks Mme Bégon, 'toujours chercher autant que je le pourrai à adoucir tout ce que je vois trop amer.'[56] Frégault, summing up his study of New France during the thirty years' peace, in a quotation from Mme Bégon's biographer, de Bonnault, remarks that 'un singulier souci de charité avait humanisé les mœurs.'[57] *Souci, humaniser, charité* – all words that apply to Mme Bégon. These terms might have made her laugh had she known of their application to her, but the thought behind them she would not have found singular, for humanising manners was her *métier*.

CATHERINE RUBINGER
Mount St. Vincent University

Notes

1 Mathurin François Adolphe de Lescure, *La Société française au dix-huitième siècle. Les femmes philosophes* (Paris: E. Dentu, 1881), p. 9.

2 Elisabeth Badinter, *Emilie, Emilie L'Ambition féminine au dix-huitième siècle* (Paris: Flammarion, 1983), p. 133.

3 E. Littré, *Dictionnaire de la Langue française* (Paris: Hachette).

4 Elisabeth Bégon, *Lettres au cher fils. Correspondance d'Elisabeth Bégon avec son gendre. 1748-1753* (Montréal: Hurturbise, 1972).

5 Charles de Mouÿ, 'Etude sur Stanislas-Auguste et Mme Geoffrin' in *Correspondance du roi Poniatowski et de Madame Geoffrin* (Genève: Slatkine Reprints, 1970), p. 52.

6 Céline Dupré, *Elisabeth Bégon. Textes choisis, présentés et annotés* (Ottawa: Fidès, 1960), p. 10.

7 Chanoine Lionel Groulx, *Histoire du Canada français depuis la découverte* (Montréal, 1969), p. 293.

8 Bégon, *Lettres*, 13 février, 1749, p. 82.

9 Ibid, 14 février, 1749, p. 83.

10 Ibid, 9 juin, 1749, p. 140.

11 Ibid, 29 janvier, 1749, p. 73.

12 Ibid, 21 mars, 1749, p. 104

13 Guy Frégault, *La civilisation de la Nouvelle-France 1713-1744* (Ottawa: Fidès, 1969), p. 122.

14 Isabel Landels, 'Correspondance de Mme Bégon,' thèse présentée à la Faculté des Lettres de l'Université Laval, 1947, p. 101.

15 Mathurin de Lescure, 'Madame du Deffand, sa vie, son salon, ses amis, ses lettres,' in *Correspondance de la Marquise du Deffand*, ed. de Lescure (Genève: Slatkine Reprints, 1971), p. xcix.

16 Louis Franquet, *Voyages et Mémoires sur le Canada* (Québec: Edition Elysée, 1974), p. 15.

17 Claude de Bonnault, 'La correspondance de Mme Bégon,' in *Rapport de l'Archiviste de la Province de Québec* (1934-35), 3.

18 P. F-X de Charlevoix, 'Mémoire sur les Colonies Françoises et Angloises de l'Amérique septentrionale' (1739), p. 72.

19 G.W. Marchand (ed.), *Voyages de Pierre Kalm dans l'Amérique septentrionale* (Mémoires de la Société historique de Montréal, 1880), p. 6.

20 Frégault, *La civilisation*, p. 216.

21 Bégon, *Lettres*, début de l'année, 1753, p. 313.

22 J.C. Bonnafoux, *Voyage au Canada de 1751 à 1761* (Québec, 1887), p. 35, cited by Frégault, *La civilisation*, p. 217.

23 *Voyages de Pierre Kalm*, p. 102.

24 Charlevoix, 'Mémoire,' cited by Frégault, *La civilisation*, p. 216

25 Dupré, *Elisabeth Bégon*, p. 15.

26 Bégon, *Lettres*, 2 juin, 1749, p. 137.

27 Franquet, *Voyages*, p. 17.

28 Mme du Deffand, letter to Walpole, 17 mai, 1767, in *Madame la Marquise de Deffand. Lettres à H. Walpole, Voltaire et quelques autres* (Paris: Plasma, 1979), p. 36.

29 Bégon, *Lettres*, 17 avril, 1749, p. 114.

30 Ibid, 13 oct., 1750, p. 219.

31 Ibid, 19 avril, 1749, p. 115.

32 Ibid, 19 fev., 1749, p. 86.

33 Ibid, 24 juillet, 1750, p. 197.

34 Ibid, 2 juin, 1749, p. 137.

35 *Voyages de Pierre Kalm*, p. 214.

36 Ibid.

37 Frégault, *La civilisation*, p. 217.

38 Franquet, *Voyages*, p. 57.

39 Frégault, *La civilisation*, p. 217.

40 Ibid.

41 Bégon, *Lettres*, 22 avril, 1749, p. 117.

42 de Mouÿ, *Correspondance*, p. 41.

43 Bégon, *Lettres*, 20 fev., 1751, p. 262.

44 Ibid, 2 juin, 1749, p. 137.

45 Ibid, fev., 1749, pp. 71-91.

46 Ibid. début de l'année, 1753, p. 313.

47 Mme du Deffand, to Walpole, 21 mai, 1766, in *Madame la Marquise du Deffand*, p. 17.

48 de Mouÿ, *Correspondance*, p. 32.

49 Bégon, *Lettres*, 20 juin, 1750, p. 192.

50 Ibid, 1 juin, 1750, p. 180.

51 Ibid, 3 mai, 1750, p. 168.

52 Ibid, 3 mai, 1752, p. 292.

53 Ibid, 23 nov., 1750, p. 239.

54 Ibid, 18 août, 1752, p. 301.

55 Dupré, *Elisabeth Bégon*, p. 11.

56 Bégon, *Lettres*, 3 mai, 1752, p. 292.

57 de Bonnault, *RAPQ*, p. 3.

7. New Trends in the Form and Function of the Dénouement: German *Empfindsamkeit* and *Sturm und Drang*

The Romantic Era in Europe has long been regarded as an age of literary change – an age in which formal conventions are undone, generic distinctions transformed, loosened, or blended, and in which the original genius of the artist creates, through the unleashing of his subjectivity, a constantly shifting mosaic of the dense interrelationships between subject and object, the self and the external world.[1] The programmatic statements of early Romantic theorists reinforce this rupture with literary conventions. Friedrich Schlegel, for instance, repeatedly maintains that the poetics of the past are both restrictive and outworn and that the contemporary age must move toward a fuller and more organic conception of art:

> Die meisten Vorstellungsarten vom poetischen Weltsystem sind noch so roh und kindisch wie die ältern vom astronomischen vor Kopernikus. Die gewöhnlichen Einteilungen der Poesie sind nur totes Fachwerk für einen beschränkten Horizont.[2]

The polemical insistence of the Romantics on literary innovation must not obscure, however, the currents of change which were already in motion in the course of the eighteenth century. Two elements of the interrelationship of literary texts must be kept in mind: first, that individual works form an interlinking chain, whereby new additions to the system

MAN AND NATURE / L'HOMME ET LA NATURE III/1984
0824-3298/84/0300-0101 $01.25/©C.S.E.C.S./S.C.E.D.S.

must be seen within the context of the whole in their dynamic interaction with other texts; and second, that literary changes are seldom arbitrary or isolated.[3] Conventions must exist before they may be broken, and literary innovation itself is frequently composed of a gradual series of changes. Transitional developments must not be overlooked, since shifts in aesthetic outlook and practice interlink with transformations in ethos and social, political and historical circumstances.

One such seemingly abrupt but, in reality, gradually evolving change may be found in the form and function of the dramatic dénouement. Although clear-cut generic distinctions between tragedy and comedy are, in the wake of the eighteenth century, no longer applicable, the present study limits itself to the 'serious' genres of drama rather than to the predominantly comic ones. My findings from an ongoing study of closure reinforce the thesis that, to varying degrees, the form and function of the dénouement of the 'serious' Romantic drama marks a radical departure from traditional closure practices in preceding dramas from Antiquity to the late eighteenth century.

Traditionally, the dénouement of the serious drama preceding Romanticism is closed, in that the fate of the major characters is determined and no major questions or problems are left unresolved. Conventionally, the drama moves from the peripety of the climax to the somewhat static equilibrium of the ending. In short, the complications of the action have been unravelled, and the closing passage of the play marks, as Aristotle states, a falling action.[4] The emotions of pity and fear reach a peak and lead to release in the cathartic action of the dénouement.[5] Not only are the survivors of the tragedy depicted in the drama led to a new harmony and equilibrium, but also the spectators share in the new vision of peace. As Theophil Spoerri summarizes, the spectator

> ... spürt, daß es eine Macht gibt, die seinem Leben einen neuen Sinn geben kann ... Für einen Augenblick fühlt er sich im Frieden mit der Welt. Es ist die eigentümliche Ruhe, die unerwartet mitten im tragischen Spiel den kämpfenden Helden überfällt und sich geheimnisvoll dem Zuschauer mitteilt.[6]

Many examples may serve to illustrate traditional closure practice. *Hamlet*, for instance, although in many ways rupturing the classical unities, employs a typically harmonious and stable outcome. As in Shakespeare's other historical tragedies, the final speakers in *Hamlet* are secondary characters who have remained more or less intact throughout the course of tragic events and who represent the forces of stability and continuity in the existing social order. Their role in the dénouement is twofold: first, to eulogise the life and death of the hero; and second, to embody the concepts of survival and restored equilibrium. Hamlet's successor Fortinbras efficiently and decisively sets about the reestablishment of a stable monarchy, as the play is brought down from the heights of pity and fear reached in the preceding scenes.

Robert Heilman comments on the significance of the conventional dénouement:

> This ending, of course, is a convention, a formalized management of proprieties that had authority in their day. Yet the convention itself is significant: it could not exist, we may conjecture, without faith in a transmittable order of values, and belief in continuity; the ending is a symbol of something earned in the tragic action, in contrast with the total sense of loss in dramas of despair and disaster. The dramatist's own sense of continuity may be more or less profound, and he may build the convention firmly into the plot or employ it casually, like a verbal tag.[7]

With the advent of Romanticism, however, the traditional forms and functions of dramatic closure are radically altered. Not only are the classical unities of time, place and action definitively shattered in Romantic theory and practice, but also the unresolved, abrupt, or turbulent dénouement is employed. Tension and emotion continue, often unmitigated, to the very end of the play; strands of unresolved conflict may remain, questions and problems may be left unanswered, and the action of the drama may come to a merely artificial halt, propelled forward, in the minds of the audience or reader, to an unknown outcome beyond the formal ending. Sudden surprise developments may transform the ending into a *coup de théâtre,* and final events frequently may not evolve rationally and inevitably from the body of the plot. Man may remain at odds with society and unreconciled with the limitations of human existence. The concept of a *heile Welt* (a benevolent world order) frequently disappears as world history and, in particular, the exceptional individual are caught up in a cycle of an ongoing universal tragedy. The dénouement may, moreover, lose its unity of focus if a wide variety of emotions and responses is depicted, and one level of perception is kaleidoscopically substituted for another. At the same time, striking visual effects may predominate over verbal elements.

These radical changes in the form and function of the dénouement culminate in Romanticism.[8] Nevertheless, the roots of dramatic change must be traced in the earlier literary period. Already in the eighteenth century, currents of change in the use of the dénouement are evident, along with a progressive approximation to the dramatic practice of the Romantics.

In what has been referred to as 'the most important play to appear in Germany before Lessing's *Miß Sara Sampson,*'[9] the dénouement is still modelled upon largely conventional lines. At the close of Gottsched's *Der sterbende Cato* (1730), Cato lies dying of a self-inflicted wound when revelation of his tragic flaw comes to him. He has, he realizes, carried his love of freedom too far, so that it has been transformed into obduracy. His uncompromising struggle with Caesar has served to undermine the stability of the Roman state. Nevertheless, the ending of the play does not flounder in despair. At the moment of Cato's death, Providence acts to reward him for his previous virtue as news is brought of the triumphant resolution of his conflict with Caesar. Cato's final words reaffirm his trust in Providence, and suggest that, through rightful conduct, man will eventually achieve his aims of a stable and enlightened existence. Addressing the gods, Cato says,

> Ihr kennt ja
> unser Herz, und prüfet die Gedanken!
> Der Beste kann ja leicht vom Tugendpfade wanken.
> Doch ihr seyd voller Huld ... [10]

What emerges from this dénouement is a traditionally harmonious vision, cloaked in Enlightenment optimism, which triumphs over tragic despair. As Heitner observes,

> Man in this tragedy is viewed, not as a lone being achieving a sad greatness in his defiance of the hostile universe, but as an honestly striving, somewhat handicapped pupil ... forgiven and rewarded in the life to come. ... In the neat and orderly universe of this play, the same general rule of reason and virtue is held up to every action so that all regrettable occurrences can be explained as the result of failure to observe the rule ... .'[11]

In spite of the death of the hero, the forces of stability and continuity in the existing social order are predominant. Equilibrium once more prevails and serves to bring down the play from the heights of pity and fear of the preceding scenes. In Aristotelian terms, catharsis is complete.

Only twenty-five years later, however, with Lessing's *Miß Sara Sampson* (1755), and even more with *Emilia Galotti* (1772), a shift away from traditional conventions, along with a corresponding change in ethos, is evident. At this point, it is necessary to look back upon the changes in dramatic theory and practice which take place in a European as well as a German context. In the course of the eighteenth century, the Aristotelian-based genre of 'tragedy' is loosened, and the situations and problems of its protagonists so far changed that, in effect, the generic designation is expanded and modified. The tragedy loses its preoccupation with socially exalted individuals and is transformed into the *tragédie bourgeoise* of Diderot or into the German *bürgerliches Trauerspiel.* From the French *comédie larmoyante* (literally, the 'tearful comedy'), the serious drama takes up a sentimental tone, aiming to engage the spectator or reader in tears of compassion. Both the sentimental mode and the bourgeois setting of tragedy find their forerunners in England with Richardson's novel *Pamela* (1740), a portrait of a supremely virtuous and tearful bourgeois heroine, and with Lillo's *London Merchant* (1731), a domestic tragedy depicting the moral and physical demise of a merchant's apprentice. Similarly, Lessing's bourgeois tragedy, *Miß Sara Sampson,* moves to the intimate milieu of the bourgeois family in a relatively humble social setting and imparts to its characters and events an emotional immediacy hitherto unknown in German tragedy.

In Lessing's bourgeois tragedy, traditional moral values are profoundly questioned. As Benno von Wiese comments, 'das Menschliche und das Moralische fallen nicht einfach zusammen, sondern durchdringen sich spannungsvoll.'[12] The plot is simple: a young girl falls in love and elopes with her profligate lover, who refuses to marry her. Sara's father, Sir William, although initially stern, pursues her, forgives her and her lover's transgressions, and begs them to return to his home. Both father and daughter admit guilt: Sara, for her lack of modesty and moral firmness, and Sir William, for his initial moral rigidity and condemnation. Lessing thus suggests that human fault stemming from innocence must be forgiven and that mercy should be present in all moral judgements.

In spite of Sara's basic virtue and her father's final generosity, their mistakes have dire consequences. The outcome does not entirely reinforce the idea of a *heile Welt.* Sara is reconciled to her father only to die of a slow poison administered by her lover's ex-mistress. The dénouement thus reveals the powerlessness of virtue and mercy in the face of misfortune and evil. Obliquely, the outcome suggests that the social order of the world is such that deceit and raw power triumph over goodness. The final atmosphere is, however, one of reconciliation as, in a final gesture of generosity and forgiveness, Sir William determines to adopt the illegitimate daughter of Sara's murderess and seducer. His closing words reflect composure and a determination to continue a constructive life:

> Laß mich nicht länger, Waitwell, bei diesem tötenden Anblicke verweilen. Ein Grab soll beide umschließen. Komm, schleunige Anstalt zu machen, und dann laß uns auf Arabellen denken. Sie sei, wer sie sei: sie ist ein Vermächtnis meiner Tochter.[13]

Greater complexity infuses the dénouement of *Emilia Galotti* and, far more than in *Miß Sara Sampson*, the foundations of the *heile Welt* are shaken. Edoardo Galotti, the model of bourgeois *Rechtschaffen*, slays his daughter Emilia – an action which she herself desires – to save her from forcible seduction at the hands of the licentious Prinz von Gunstalla. Private morality and the corrupt power of the state have thus come into a conflict which can be resolved honourably only through the self-sacrifice of virtue. Galotti's desperate upholding of virtue ironically transforms him into a murderer in the eyes of the state; at the same time, his earthly judge, the corrupt prince, is a tyrant who heeds no moral or social laws. The dénouement offers no immediate solution to this dichotomy between individual morality and a corrupt society. At the same time, Galotti is convinced of final justice at the hands of God: 'Ich gehe und liefere mich selbst in das Gefängnis. Ich gehe und erwarte Sie als Richter. – Und dann dort – erwarte ich Sie vor dem Richter unser aller!'[14] This justice is, however, only partial, removed from the actual world. No solution is offered for the mockery of justice and the abuse of power upon earth. The equilibrium of the dénouement is therefore already beginning to become unbalanced in eighteenth-century German drama.

Analogous dénouements exist in many dramas of the *Sturm und Drang* period in the later eighteenth century. Behind the adolescent furor of *Sturm und Drang*, the melodramatic posturings of its rebellious heroes, its moral relativism, and the runaway eloquence of its language lies much, in fact, which links it with *Empfindsamkeit*. The family is still much in the forefront, although it has now been infused with internal conflict; social criticism, rendered sharper by the iconoclastic tendencies of the *Sturm und Drang* hero, continues, and has now expanded to take on a historical dimension; and sentimentalism has led to bombastic language and unbounded bursts of emotion. At the same time, the social and political tensions depicted in these dramas still do not lead to a nihilistic rejection of society or moral standards. Equilibrium still partially operates in the dénouement.

The *Empfindsamkeit* and *Sturm und Drang* dramas included in this study share one common element in their closure: all do indeed represent a resolution of the action. Following Aristotle's prescription, they represent a complete action, and, unlike many Romantic dramas, do not leave the audience or reader in suspense as to the eventual outcome. With the possible exception of Wagner's *Kindsmörderin* (1777), the German eighteenth century drama is complete with the dénouement. In *Die Kindsmörderin*, the execution of the seduced heroine who has slain her illegitimate child is imminent, while her repentant lover, v. Gröningseck, is left in paroxysms of his 'Reue nach der Tat.'[15] Nevertheless, the apparent desertion of Evchen by her seducer is revealed as a gigantic mistake resulting from the perfidy of a deceitful friend, and some hope remains for a royal pardon as v. Gröningseck determines to rush to Versailles to put his case before the king:

Und Ihnen nebst Ihrer ganzen kriminalischen Unfühlbarkeit zum Trotz, mein Herr! will ich mich heut noch auf den Weg nach Versailles machen, bei der gesetzgebenden Macht selbst Gnade für sie auszuwürken, oder — [16]

The dénouements of *Empfindsamkeit* and *Sturm und Drang* dramas fall into two major groups and one sub-group. First, there are those of reconciliation which reveal a partial resolution of tensions, exemplified by *Miß Sara Sampson*, Goethe's *Götz von Berlichingen* (1773), Leisewitz's *Julius von Tarent* (1776), and Schiller's *Die Räuber* (1781). A subdivision of this group is formed by dramas which employ the distancing devices of irony in the dénouement. Into this group, the 'reconciliatory-ironic,' fit Lessing's *Emilia Galotti*, Lenz's *Die Soldaten* (1779), and Schiller's *Die Verschwörung des Fiesco zu Genua* (1783). The second major division I have chosen to call the dénouements of 'sustained tension.' On the one hand, their closing scenes follow naturally and fairly predictably from the body of the plot, and serve to complete the action, leaving few major problems unresolved; on the other, they end explosively, with little of the falling-off of tension or emotion revealed in the first major group. Representative of this group are Gerstenberg's *Ugolino* (1768), Klinger's *Zwillinge*, and Wagner's *Reue nach der Tat* (1775). An even more transitional dénouement with strong elements of the frequent open-endedness of Romantic drama is, as observed above, that of Wagner's *Kindsmörderin*.

The outcome of *Miß Sara Sampson*, our first example from the 'reconciliation' endings, employs, as we have pointed out, a 'falling action,' as the resigned composure and determination of Sir William to continue with the matters of living lead to a relaxation of emotion and renewed stability. To a smaller degree, Goethe's *Götz von Berlichingen* concludes on a note of elegiac calm. Reconciliation takes the form of resignation to a corrupt age and a desperate hope for social reform in the future. In the present age, however, harmony is found only beyond the earth: 'Nur droben, droben bei dir,'[17] i.e., with God in the afterlife. On another level, Karl Moor in *Die Räuber*, an angry young man who has countered the corruption of society with lawlessness and violence, performs an about-face in the dénouement of the drama. After the loss of his entire family, including his father and his lover, he comes to the tragic realization that to attempt to change the world order by repaying evil with evil merely propagates destruction.[18] Only by giving himself up to justice, he determines, can he achieve forgiveness for his sins. Through his imminent death, ironically, he reenters the human society he has previously spurned. The final reconciliation is, however, only partial. The antagonism of the exceptional individual and a restrictive social milieu is only artificially resolved through the hero's repentance, and the final harmony is purchased at the price of total defeat and self-destruction.

While notes of triumph are heard in the reconciliatory group, the reconciliatory-ironic dénouement contains little of this positive element. In spite of the falling-off of action and emotion, irony injects a discordant note. Schiller's *Verschwörung des Fiesco zu Genua*, for instance, ends with the defeat of the republican cause in Genoa, spurred on by the *superbia* of Fiesco, the erstwhile revolutionary leader. Only one action, the removal of the new tyrant and the restoration of the previous but more benevolent tyrant, can save the state from disintegration. The understated words of Verrina, the self-appointed murderer of Fiesco, ironically emphasize the failure of the revolution: 'Ich geh zum Andreas.'[19] Revolution has thus accomplished a complete circle which leaves political conditions finally unchanged, and the reinstatement of equilibrium takes the form of compromise. One other innovative technique is employed in the stage directions: *'Alle bleiben in starren Gruppen stehn. Der Vorhang fällt.'* With the freezing of stage movement in a tableau of dumbfounded spectators, Schiller precipitates a prominent device of the later Romantics. Not only words but also the suspension of symbolic and significant action brings the dénouement to a new prominence. Similarly, the closure of Lenz's *Die Soldaten* ironically emphasizes the circularity and lack of resolution in human society. The dénouement only restores harmony insofar as father and daughter are reconciled and reunited; in a type of double dénouement, the closing discussion between the pseudo-enlightened Gräfin and Obrister on the problems of soldier morality throws ironic light upon the ongoing corruption and blindness of society.

The second major group of dénouements, those of 'sustained tension,' offer little falling-off of emotion and hence do not exemplify traditional catharsis. One of the earliest *Sturm und Drang* dramas, Gerstenberg's *Ugolino*, concludes upon a note of renewed harmony which is, however, far removed from the subdued calm of the traditional dénouement. To the accompaniment of a crescendo of background music, the hero determines with courage to face inevitable death by starvation. His final words are emotionally charged and endow a climactic effect upon the dénouement: 'Ich hebe mein Auge zu Gott auf. Meine zerißne Seele ist geheilt ... Und dann seid mir gepriesen, die ihr diesen Leib der Verwesung hinwarft! Ganz nahe bin ich am Ziel.'[20] A more open-ended dénouement is found in Klinger's *Zwillinge.* Although the action is, in one sense, resolved through the father's punishment of his younger son's murder of the elder brother, the play ends in an explosion of despair as the father cries, 'Rächen will ich Vater Guelfos Sohn! erretten von der Schande Guelfos Sohn! leben im Jammer verwaist —'[21] and stabs his son. Through the slaying of his own younger son to avenge the murder of the elder, he has, in effect, rendered himself childless. This self-destructive act of justice is an admission by Guelfo of his failure to accomplish the ideals of love and family loyalty which have formed the basis of his life. As his final words intimate, his future life will be devoid of harmony, a time of prolonged suffering.

In conclusion, the dénouements here outlined are significant, both in themselves and in their literary-historical position as indicators of a major change in dramatic theory and practice. They throw light upon the *Weltanschauung* of the later eighteenth century and upon the innovative nature of an age breaking away, philosophically and artistically, from traditional forms and attitudes. While *Der sterbende Cato,* our example of a traditional outcome, concludes smugly with a belief in the best of all possible worlds and the perfectibility of man through virtue and reason, the dramas of *Empfindsamkeit* and *Sturm und Drang* reveal a gradual loss of belief in the foundations of a *heile Welt.* As precursors of Romantic dramatic practice, they already point to the open-endedness, lack of resolution, ambivalence, and melodramatic explosiveness of Romantic endings.

The first type of dénouement here outlined — the 'reconciliatory,' with its subgroup, the 'reconciliatory-ironic,' is transformed into three basic types in Romanticism. First, the 'transcendent' dénouement of the Romantics resolves existential dilemmas by rejecting immediate reality in favour of a higher, transcending vision. The inverse of this type, of which the 'reconciliatory-ironic' is the precursor, is found in the 'nihilistic' dénouement of Romantic dramas, where no resolution is found, and the protagonists continue to be the prey of destructive forces. Yet another group in Romanticism, the 'circular' dénouement, is closely allied to the previous one. It contains no positive resolution for, although the fate of the protagonists is outwardly determined, humanity continues to play the part of a tragic (or even tragi-comic) hero in an ongoing world tragedy. Finally, the second group of *Empfindsamkeit* and *Sturm und Drang* dramas, those of 'sustained tension,' transforms itself into two types in Romanticism: the 'climactic' dénouement and the 'symbolic tableau.' The climactic dénouement replaces the traditional falling action of the conclusion with a rising one, and frequently leaves questions unanswered and the action only partially resolved. A *coup de théâtre*, or surprise twist in events, is often incorporated. The 'symbolic tableau' subordinates verbal to visual elements. The final events, unresolved tensions, and even the future course of action are immobilized and held in a concluding 'stage picture' or tableau, upon which the curtain falls.

German drama, it must be remembered, does not move exclusively from *Empfindsamkeit* and *Sturm und Drang* to Romanticism. The 'angry young man' movement of the *Sturm und Drang* was short-lived. Its two most prominent survivors, Goethe and Schiller, shifted in their later years to a Classical ideal, and reveal, in their later dramas, little of the open-endedness of their *Sturm and Drang* work. Nevertheless, the innovative trends introduced in the latter half of the eighteenth century are further developed and expanded by the new generation of Romantic dramatists at the opening of the nineteenth century. The radical changes accomplished in the form and function of the dénouement of serious drama become a characteristic and distinguishing feature of European Romantic drama. Already in the eighteenth century, however, the dénouement emerges as a significant indicator of a new *Weltanschauung* and serves as a vantage point from which we can examine the emergence of a new ethos and aesthetic practice. Most important, the study of changes in the form and function of the dénouement illustrates the dynamic linkage of innovative trends in literary history.

ANNA WITTMANN
University of Alberta

Notes

1 See H.H.H. Remak, 'West European Romanticism: Definition and Scope,' in *Comparative Literature: Method and Perspective*, ed. N. Stallknecht and Horst Frey (Carbondale: Southern Illinois University Press, 1971), pp. 275-311.

2 *Werke in zwei Bänden*, ed. Wolfgang Hecht (Berlin and Weimar: Aufbau Verlag, 1980), I, 'Fragmente,' No. 434, p. 255.

3 See T. Todorov, *The Fantastic. A Structural Approach to a Literary Genre*, trans. Richard Howard (Cleveland: Case Western Reserve University Press, 1973), p. 2; also Maria Corti, *An Introduction to Literary Semiotics*, trans. D. Mandelbaum, (Bloomington: Indiana University Press, 1978), p. 2: 'Literature is not the sum of its texts but a kind of totality linked and linking, and in movement.'

4 Aristotle, *Poetics,* trans. and ed. Gerald F. Else (Ann Arbor: University of Michigan Press, 1967), p. 36.

5 Ibid., p. 36.

6 Theophil Spoerri, 'Das Problem des Tragischen. Aus der Perspektive des französischen Dramas gesehen,' in *Tragik und Tragödie,* ed. V. Sander (Darmstadt: Wissenschaftliche Buchgesellschaft, 1971), p. 78.

7 *Tragedy and Melodrama: Versions of Experience.* (Seattle: University of Washington Press, 1968), pp. 160-61.

8 These findings are based on my dissertation project, 'The Structure and Function of the Dénouement in German, French and English Romantic Drama,' currently in progress. An abbreviated version is due to appear in the International Comparative Literature Association's forthcoming *History of Romantic Drama in European Languages,* ed. G. Gillespie.

9 Robert Heitner, *German Tragedy in the Age of Enlightenment* (Berkeley: University of California Press, 1963), p. 26; see also Jürgen Jacobs, 'Daß klassizistische Drama der Frühaufklärung' in *Handbuch des deutschen Dramas,* ed. Walter Hinck (Düsseldorf: Bagel, 1980), p. 63.

10 Johann Christoph Gottsched, *Ausgewählte Werke,* ed. Joachim Birke (Berlin: De Gruyter 1970), II, p. 113.

11 Heitner, p. 31.

12 Benno von Wiese, *Die deutsche Tragödie von Lessing bis Hebbel* (Hamburg: Hoffmann und Campe, 1961), p. 33.

13 Gotthold Ephraim Lessing, *Gesammelte Werke,* ed. Wolfgang Stammler (München: Hanser, 1959), I, p. 398.

14 Ibid., I, p. 624.

15 The prominence of the theme of remorse in Wagner's work is revealed by the title of an earlier drama, *Die Reue nach der Tat* (1775).

16 Heinrich Leopold Wagner, *Die Kindsmörderin* in *Sturm und Drang: Eine Auswahl,* ed. Erich Loewenthal and Lambert Schneider, 3rd ed. (Heidelberg: Schneider, 1972), II, p. 605.

17 Johann Wolfgang von Goethe, *Götz von Berlichingen* in his *Werke,* ed. Wolfgang Kayser, IV (Hamburg, 1953), p. 175.

18 Friedrich Schiller, *Sämtliche Werke,* ed. Gerhard Fricke et al. (München: Hanser, 1962), I, p. 617.

19 Ibid., p. 751.

20 Heinrich Wilhelm von Gerstenberg, *Ugolino* in *Sturm und Drang: Eine Auswahl,* I, p. 61.

21 Friedrich Maximilian Klinger, *Die Zwillinge* in *Sturm und Drang: Eine Auswahl,* II, p. 123.

8. Canonic Art: Pregnant Dilemmas in the Theory and Practice of Anton Raphael Mengs

The problem of canonic art recurs in history. Whenever artists and aesthetes assume leadership because they are dissatisfied with the state of artistry, are intent upon significantly changing a reigning style, or feel called upon to defend the status quo, such leaders in the artworld exercise certain preferences, lending their authority to their choices. The choices made by leading artists, theorists, critics, and even art patrons, have a way of assuring that the ensuing art is normative, a canon for subsequent art. The dilemmas concealed in this establishment of canonic art surface with special clarity and historical importance in the work of Anton Raphael Mengs (1728-79).

Mengs was a published art theorist to whom the renowned Winckelmann dedicated *Geschichte der Kunst des Altertums* in 1764. Mengs was also a celebrated artist. In 1751 he was named principal painter at the Dresden court of Augustus III, King of Poland. In 1759 he was commissioned in Rome by Cardinal Albani to do the crowning fresco of the *galleria nobile* of the whole villa complex which served as salon for papal entourages and for illustrious visitors from foreign lands. In 1763 Mengs was instated as Director of San Fernando to establish a Spanish national Academy of Art. He became the principal painter at the Madrid court of Charles III in 1766. Pupils tutored by Mengs and younger kindred spirits became directors of art academies and their curricula in Turin, Stuttgart, Dresden, Karlsruhe, Copenhagen and Vienna.[1] Mengs was also a profes-

MAN AND NATURE / L'HOMME ET LA NATURE III/1984
0824-3298/84/0300-0113 $01.25/©C.S.E.C.S./S.C.E.D.S.

sional art historian who both assisted and corrected Winckelmann, perhaps even spoofed him, in the analysis of Graeco-Roman art discoveries.[2] Mengs' method of art criticism supplied Goethe and Heinrich Meyer with what they needed to give their classicist Weimar *Kunstfreunde* direction.[3]

That is, Mengs practised an uncommon range of artistic tasks, and had the freedom of position – especially as a gifted outsider asked to come into a cultural situation – to formulate his ideas about artistic excellence. Mengs was able to define a coherent praxis and policy for art. His tenets are worth analysis, especially if we recognize that the problem of giving leadership in painting is a matter of vital importance for the art-world of any day. Are there ways to posit artistic norms, canons for art, that will be neither dogmatic nor *laissez-faire?* Can the conception of canonic art be fruitful rather than abortive?

The six theses of Mengs' art theory

Gedanken über die Schönheit und über den Geschmack in der Malerei took shape during conversations with Winckelmann in Rome and was finally published in 1762 in Zürich by the father of Henry Füssli. The core doctrines of Mengs' theory are clear: (1) There is an invisible, absolute, divine Perfection in which each form of Nature participates, according to its specific, graduated destiny (1:9, 13, 15).[4] (2) Anything which visibly conforms to our idea of its telos is considered beautiful, because Beauty is visible Perfection (1:7-8, 10-12, 17, 21). (3) Since some things are more beautiful than other things, good taste is the rational selection of the most beautiful forms of Nature (1:26-30). (4) The art of painting, like poetry and music, can surpass Nature in Beauty, 'meliorate the things in Nature,' because the artist, like the bee extracting sweetness from many flowers to make its honey, can freely choose to imitate the most beautiful parts found here and there in Nature for composing the artwork (1:18-19, 30, 51).

In the important treatise of 1776 addressed to Don Antonio Ponz, secretary of the Royal Academy, San Fernando, Mengs takes a stand on matters then current which gives his theory of canonic art its cachet: (5) This painterly imitation of what is essential in the natural objects perceived – using design, chiaroscuro and colour – should be ideal, and will be 'the more valuable as the idea conveyed will be perfect, distinct, and clear' (P, 10-13).[5] (6) Grecian artists have excelled in perfecting the over-all rendition of Beauty; but even the Greeks would be astonished at the admirable precision of expression in Raphael's design (P. 52, 106), and although Correggio is not so great as Raphael in painting mental states, Correggio painted bodies in chiaroscuro so enchantingly he 'completed' the art of painting, and 'was the meridian of the art: from that point it always went declining' (3:53); and Titian is the master of true colouring. These three artists in their respective strengths are the touchstones for painterly art and the best taste (1:20, 37-38, 40, 78; P, 44-45). By way of digression, it should be mentioned that Mengs elaborates in theory what he preaches for artistic praxis; it is unfair to brand his amalgam of ideas eclectic or muddled.[6]

Certainly, Mengs' description of the five principal parts of painting in which 'invention' and the 'composition' subservient to it form the principles constituted by design, chiaroscuro, and colouring (P, 36-43, 98-103; 3:96-97) repeats what had become a commonplace among European artists and thinkers since the writings of Vasari and Dolce, Lomazzo and Zuccaro.[7] Mengs' metaphysic of Beauty, trailing thin Meso-Platonic clouds of deity in Nature which our rational soul inspects to find what comports there with our human idea of the more and most beautiful (1:22-23, 30), articulates precisely the major thesis of Batteux's *Les beaux arts réduits à un même principe* (1746), namely, 'l'imitation de la belle nature.'

The fact that Batteux's thought was in vogue among Germans after Schlegel's 1751 translation,[8] the fact that the writing of Mengs constantly echoes Bellori and the Italian art theorists of cinquecento,[9] and the fact that Lomazzo was Mengs' favourite reading[10] are all evidence that Mengs was steeped in tradition. His theory of beauty, taste, and art, far from being a pastiche of odd thoughts, is intellectually coherent. His theory even conforms to the age-old epigram of Zeuxis, updated: cross expression from Raphael's design, harmony from Correggio's chiaroscuro, and true representation from Titian's colour, and moderate your inventive composition with the Greek spirit of reserved perfection, and you have the *summum bonum* of Beauty, taste, and painterly art (1:37-41 & 1:part III *passim*).

Whether Mengs appropriated the Zeuxis fable from Bellori or (more probably) from Lomazzo, Alberti, and Pliny, or first heard of it in conversation with Winckelmann[11] neither matters nor detracts from his genius in having turned it into a theory of art tied to a practical norm of visible Beauty and taste with a canon for painting that could be known, taught, and was believed to be important for the state of what Pico della Mirandola earlier called *De hominis dignitate* (1:33). And Mengs put these 'Thoughts' down in writing which was accessible to non-specialists, in short chapters that should have warmed any popularizing encyclopedist's heart. His *Gedanken* antedated Schiller's short series of letters *Über die Ästhetische Erziehung des Menschen* (1794-95) by a generation. Nothing systematic like Mengs' *Gedanken* had appeared in German before 1762. No wonder it was acclaimed and went through five editions before his death.[12]

A close scrutiny of the texts corrects the unfavourable Romantic reading given them and might end superficial characterization of his theses about canonic art.

Despite the Meso-Platonic flourishes of his doctrine, Mengs' basic approach is not speculative, but is Aristotelian.[13] His theory of imitation takes its cue from Aristotle's conception of *mimesis,* according to which the artist seeks to re-present 'the motive of its model.' Mengs explicitly contrasts such 'imitation' with Plato's concept of the journeymanlike copy (1:35, 44;2:105; P,10-12, 112-116). The artist-theorist Mengs also very matter-of-factly gives precedence to training of the eye and hand over learning rules for art. Extrapolating from his own history, Mengs unequivocally posits that practice and execution should precede theory and scientific rules in the education of an artist (1:3). Such are not the dicta of a scholastic, arm-chair theorist.

While Mengs shared with Winckelmann a connoisseur's veneration for the art of antiquity, Füssli was wrong to lump them both together as *antiquomanes.* Mengs was a conservationist in theory, but he valued the Greeks not because they were the faultless paradigms Winckelmann somewhat indiscriminately believed them to be, but because the Grecian painters, by moderating their artistic attempts to cover equally well all the necessary parts of imitating 'the whole of Nature,' had fashioned the best examples of complete Beauty (1:38-39, 75-77). For Mengs the paragons of painting are Raphael, Correggio, and Titian. That fact helps distinguish what Mengs is doing in 1762 from the antiquarian concerns of Winckelmann but also from the habit of latching onto 'Greek connections' prevalent with a Mannerist thinker like Junius (1512-75) or with Bellori (1615-96).

The distinctions are nice, but such fine distinctions are important if art historians are to assess accurately what is historically particular in Mengs' 'neoclassical' thinking. Cinquecento theorists used the rhetoricians of antiquity to bolster painting's claim to be as honourable as poetry, and did it often rather pedantically. Seicento thinkers on painting were not concerned with an apology for painting so much as justifying *la grande manière* with an aura of ancient themes and classical detail (cf Poussin) and reinterpreting Graeco-Roman theories with an academic accent. Mengs, however, simply includes, and thereby relativizes, the Graeco-Roman statuary, the tales of Greek painting (and the findings at Herculaneum after 1738), as Exhibits A, so to speak, in his pantheon of canaonic art which is filled with *other* exhibits.

Although he would whistle the allegro of a Corelli sonata while painting an 'Annunciation' and switch to a melody of Dorian mode for 'Deposition from the cross,'[14] and although he continually drops phrases, for example, in describing 'invention' as 'the essence and poetry of a picture' (P,99-100), Mengs is not at all concerned with arguing about *ut pictura poesis* or the 'correspondences' between painting and music. That vocabulary is simply the idiom in which Mengs focuses all his attention upon bringing an encyclopedic theory of canonic art ('*all* those parts' of painting, '*all* the rules of the art' [my italics], 1:3) into play *practically*, for beginning, intermediate, and professional painters (1:4). Mengs' theory is 'neoclassical' only in the sense that it reveals a belief in ancient fundamentals (2:145-46), the 'tried-and-true' ideas and methods accumulated throughout the ages which he trusts will reintroduce an enlightened understanding and performance to the art of painting.

There are two dilemmas posed by Mengs's theory of canonic art that need particular attention, one stemming from the systematic structure of his aesthetic theory (theses 1-4), and a second from his neoclassical temper (theses 5-6).

Firstly, if the best Beauty the art of painting achieves is one of supernatural simplicity[15] (both by concentrating natural beauties and by neglecting encumbering minutiae [2:113, 116;P,11]), will the best painterly art tend to leave actual 'Nature' behind and use objects only to reflect Beauty?[16] And if the most noble part of painting is to rouse (but not ruffle!) our 'generous passions' and to instruct the human mind (2:95; P,9-11,90), must painters reinvest the normative geometric uniformities and perfect primary colours, – upon whose infinite variety painterly beauty depends (1:10-13), – with 'poetic' ideas, or not?[17]

Secondly, if the art of painting attains the most sublime degree of beauty and perfection the ancient Greeks gave it (1:20-21), is the only possible historical change one of decline (2:117)? Or, if painting is always more or less perfect (cf 'gradations of Beauty,' 1:15), what makes the 'rise, progress, and decay' of canonic art possible as a pattern? And if any human act, which is imperfect, can be 'carried to a higher degree' of perfection (1:24,80), why is it doubtful that another Athens might appear in Italy (1:21)? – can Athens not repeat itself?

That is, surrounded by the legitimation of decorative, easy-going art ('adapted to the discernment of those who employ us,' writes Mengs in a wry note), and aware of the ongoing loss of a literary tradition in the art of painting, Mengs unfolds a theory which combines a secularized theology of Beauty with a humanist philosophy of history. He proposes as canonic art the kind of painting which imitates 'Nature' – half-understood to be 'the best of the Western tradition of art.' Although

Mengs slips into this position more by way of practical, pedagogical concerns than of carefully argued, philosophical reasons (1:40, 78), the problematic import for theory of art is still serious: canonic art becomes essentially a kind of palimpsest.

The Parnassus Fresco of Apollo, Mnemosyne, and the Nine Muses: Villa Albani (now Torlonia), Rome (courtesy of The Mansell Collection, London)

Evidence for this reading of Mengs' theory and its dilemmas is at hand in the *Parnassus* fresco he painted in the villa Albani (1760-61). The exhaustive detective work by Steffi Röttgen on *Parnassus* with its flanking *tondi* and an earlier drawing that shows compositional change in the final painting, which she has put in context by examining all relevant treatments of the *topos*,[18] has iconographically unpacked every graphic hint in the piece. Most important is this: the god Apollo surrounded by the nine Muses prepares for the coronation of Albani as the reigning seigneur of the new Mount Olympus found at 'eternal' Rome, and gestures in presentation toward the most favoured position of Calliope, muse of painting, who leans pensively on a Doric column, signifying ancient art, and holds a scroll inscribed with the name of Mengs. Mnemosyne, mother of the Muses, seated on Apollo's right, hand to ear as prescribed by Ripa, and pointing to the spring of fresh water at Apollo's feet, emblematizes mnemonics, the *art* of remembering, which, in a tradition running from Cicero to Lomazzo and onward, was an art assigned especially to painting. Painting makes abstract ideas concretely vivid and memorable; painting functions *par excellence* commemoratively, making the past present.[19] So, like a mirror mirroring a mirror, Mengs paints the meaning of painting, that is, the best art for making present idea-filled Greek art in all its excellence.

The whole conception of Cardinal Albani, counseled by Winckelmann, was to effect a genuine restoration of Greek art in a natural setting that would give the lie to everything artificial. Every piece of authentic ancient art the Cardinal owned would be housed so as not to seem antique. Albani did not want his villa to be a museum, a showcase of replicas,[20] but to recapture the glory of a real ancient Roman villa alive with Greek art. Every artwork must not seem but *be* Greek. Because there was no genuinely Greek painting among the Cardinal's fabulous cache of sculptures, marble columns, bronzes, mosaics, dishware, reliefs and valuables, Mengs was commissioned in 1759 to produce a Greek painting that would both celebrate this new centre of Greek art and *be* its climax.

Mengs fulfilled the commission according to the specifications of Winckelmann, who had formed strong opinions from his literary sources about what Greek painting must have been like. Two criteria are evident: one, the painting be a repository of images replete with poetically fashioned, universal ideas (*algemeine Begriffe*) ranging from mythology to the secret wisdom of many peoples, the more exquisite the allegorical dress the better;[21] two, Greek sculpture should be the paradigm for the painterly composition.[22]

This last criterion, which complements Winckelmann's remythologizing, literary concern, is of historical importance because it explains the feature which startled contemporaries of *Parnassus* into admiration and made them believe a new artistic era was dawning. There was precedent enough for the vertical perspective and non-illusionary space in *quadri riportati* done by Annibale Carracci, even Pietro da Cortona, Guido Reni and others who had decorated ceilings in *gallerie* of Roman *palazzi.*[23] Mengs certainly needed to follow that local taste and could not allow the figures above the spectators to soar off into a baroque heaven when the whole point of the commission was that the Muses have taken residence here on earth! But the sculptural feel in the painterly line provoked eyes to see something new. The contour of each figure poised like a solitary silhouette, separated from the others, and tastefully arranged in an oval tableau struck the viewer surfeited on *composizioni macchinose* as singularly different, an incredibly uncluttered, pure, restful presentation of truly living figures, statuary paragons of immortal, gracious beauty.

Parnassus is not the work of a neophyte who mimicked the Raphael piece in the Vatican, ended up with an artificial collection of mannered individuals who lack organic unity, and was praised excessively by his famous friend Winckelmann. This is the Romantic misreading which has become largely the official one.[24] Mengs knew precisely what he was doing and did it masterfully: painting a *Greek* painting! This entailed following Winckelmann's sculptural prescription (which certainly took precedence over 'copying' any *Roman* remains found at Herculaneum).[25] Mengs assumed that the Greeks bestowed great assiduity on single objects and that their most celebrated paintings and statues 'did not form one grand unity, but only an assemblage of figures, which had their particular excellence' (P, 45). Further, Mengs believed Raphael, whose genius penetrated the truly beautiful style of the ancients, was held back by the ignorance of his own times about Greek customs and ideals; Raphael's own genius 'could not entirely abandon humanity, with the happy success of the ancient Greek painters' who ranged freely 'between the heavens and the earth' (1:49,62;2:101).

Mengs' *Parnassus* was artistically premeditated; it is a principled manifesto for the remaking of Greek painting, which yet is, palimpsestically, thoroughly pro-Raphaelite. Unlike Raphael, Mengs had seen the Apollo Belvedere; so Mengs' Apollo stands naked. Mengs' improved knowledge of Greek ideals, he believed, permitted him to perfect Raphael, that is, return Raphael's genial art to a more unearthly, Greek divine Beauty.

A brief remark en passant can tie the analysis so far into a larger problem. One could do worse than remember Mengs' fresco *Parnassus* as touchstone for a graphic definition of what 'neoclassical' can mean as an art historical term for capturing precisely the cultural spirit of a time which formed by radiation, as it were, a kindred number of artists and writers who constitute a fairly definite and roughly datable period or 'movement' in Western cultural history. Whether certain traits – sculptural fixity, powerful linear clarity, serious simplicity – can serve as neoclassical differentia in other arts than painting, and in other painterly artists, for identifying this homogeneous drive to redo Greek (or Roman) art and be 'neo-classical,' will depend upon comparative examination. But there is a method of art historiography that allows for a kind of flexible precision in determining such categories. It can be illustrated in outlining a way to understand Mengs' own painterly development (even though painting by painting analysis cannot be done in this brief article).

Most commentators agree that the *Parnassus* is quite unlike the rest of Mengs' *œuvre.*[26] Another major problem for interpretation is to sort out the effects of Mengs' double-minded appreciation of Raphael and Correggio, and of his changes in locale, e.g., his moving from Dresden (dominated by Ismael Mengs) to Rome (friend of Winckelmann, rival of Batoni) to Madrid (fierce competition with Tiepolo and Giaquinto), back to Rome, then to Madrid, and again back to Rome.

Although everyone's professional development is a labyrinth of influences, and although every historian mints his own coin of terms, one way to order and understand the stylistic changes in Mengs' painterly art is this (cf schematic note): From 1744-58 Mengs was 'a promising young painter in the Raphael-Maratta tradition,' as Ellis Waterhouse puts it,[27] who was steeped in a Germanic-courtly rococo spirit.[28] Around 1759-61 with *Parnassus*[29] Mengs' love for Raphael's balanced order and measured use of colour became reinforced by a commitment to neoclassical sobriety which breathed simplification and a new idealism into Raphael's artistic features. Then, from 1762-79, in Mengs' art a submerged predilection for Correggio rose to the surface. A preference for scumbled colour and a warmer grace moderated but perpetuated the neoclassical temper.[30]

An advantage to this kind of analysis for tracking trends and shifts – if it stands verification – is the recognition that neo-classical art, epitomized perhaps by the spirit of Mengs's *Parnassus,* does not need to follow Raphael's paradigm. The neoclassical allegiance – to *be* the classical ideal *anew* – can follow Correggio too. There are various types of neoclassical painting.[31]

The pregnant dilemma posed by Mengs' *Parnassus* for painting is this: how can there be an avantgarde palimpsest, a reconstruction meant to be *schematic note on shifts in Mengs' style*

la grande manière	rococo c. 1715-75	neoclassical c. 1755-1815	romantic
[Raphael]	Mengs I (1744-58)	*Parnassus* (1759-61)	
[Correggio]		Mengs II (1762-79)	

new, painterly reform by return? If one discounts the caterwauling of 'pseudo-classicism' indulged in by critics who have an ideological axe to grind, the dilemma still remains: is repristination of art possible without committing the historical error of becoming anachronistic? How can what is historically past be made the norm for the future without becoming utopian, historically unreal? Can reiteration of a perfect canon be made the norm for art without spawning work that will be, at inception, stale? What is the legacy of the spirit of *Parnassus?*

Mengs' Framework for Writing Art History

The dilemma in Mengs' *theory* of canonic art, which confronts an artist with the task to simplify inventive composition if he wishes to reach a higher degree of Beauty, produces a curious tolerance for both artistry and taste, together with a hierarchy of Beauty. The dilemma in Mengs' *praxis* of canonic art, *Parnassus,* which purports to redo perfections of the past, prompts the emulation of the Masters of perfect Beauty, especially if divine Beauty is a teachable ideal in the Academy and can be learned through practice.

The results of Mengs' dilemmas are (1) an ontological breadth to the canon of artistic Beauty that supports Enlightenment tolerance, and (2) a Zeuxian openness to pollinate rosebuds wherever ye discriminately may. When one joins the results of these dilemmas to the artist-theorist Mengs' strong conviction that the artistic elements – design, chiaroscuro and colouring – are central to the purpose and meaning of a painting, then the upshot is an innovative, systematic methodology for 'modern' art historiography, arguably the first.

Every painter in the polite arts, writes Mengs, has imitated Nature; the only difference is that they have done it in different ways (1:75). The styles in which painters execute the 'picturesque appearance' of their ideas 'are infinite, but the principal ones from which the others should flow, may be confined to a small number' (P,14-15), five good ones, to be exact, and two unworthy 'manners.'

(1) *Sublime* style transcends in form any possible perfection found in nature; an abstracted, austere simplicity is its clue (P,17-19). (2) *Beautiful* style is distinguished by imaging every perfection possible, neither slighting nor exceeding any essential quality proper to what is imitated, in its totality (P, 19-23). (3) *Graceful* style highlights the gentleness of a modulated chiaroscuro, whose excellent agreeableness is not to be belaboured (P, 23-27). (4) *Expressive* style gives precedence to design, which exhibits by its conforming line the exact, corresponding character of what is represented (P, 27-29, 106-7). (5) *Natural* style is particularly attached to what is casual and amusing; its *forte* is the colouring of aerial perspective so requisite for being true to everyday life (P, 29-32).

Some painters, writes Mengs, have made a habit of merely affecting the virtues of other styles, thereby corrupting them. This mannered approach could be called the (A) *Vitiated* style (P, 33-35).[32] And painters who potboil artworks for the easy return of popularity may be called exponents of the (B) *Facile* style (P, 35-36).

Mengs has constituted, therefore, without much prejudice, a rough taxonomy of basic styles which provides systematic ordering for art-critical perception. Compared to Diderot's salon reports (begun in 1759) where, with an antipathy for rule-book judgments, Diderot gives voice to the sentiments welling up in him while his connoisseur's eye roves over the painting, Mengs' approach seems as detached and object-oriented as a professional. Compared to Vasari's *Vite* (1550) where, says Mengs' first editor Joseph Nicolas de Azara with a touch of spleen, the anecdotes often 'treat of everything except what is essential in the Arts,'[33] Mengs focuses attention rigorously on artistic phenomena. Because Mengs' typology of styles is based upon categories which deal in the hegemony of certain artistic features, he can credit each style with equal artistic canonicity, even though each is more equal than the others as far as its own peculiar patterning and contribution to Beauty goes.[34]

A strength of Mengs' categorical framework for analysis of paintings and sculpture is that it has a comparative scope: the framework can't help but relate and group, weight and contrast, for example, Titian, Velasquez and Rembrandt, who manifest a Natural style, next to Annibale Carracci's male figures, Guido Reni's female portraits, the Venus dei Medici and the torso of Apollo, whose lines, says Mengs, have the

Beautiful style. A person determines the style of art by examining which artistic element gives the distinctive feature to a piece. This kind of taxonomy in grappling with the art of painting in the 1760s-70s is a genial, not to say original, step on Mengs' part, toward formation of a *bona fide* art historical method that had been lacking.[35]

Further, Mengs carried his systematic sense of pattern and concern for canonic art over into the problem of temporal succession — the 'History of taste' (1:32-39), the 'Rise, progress, and decay of the art of designing' (2:110-153), and had the uncommon temerity to say:

> Painting has partaken of the same vicissitudes and revolutions as all other sublunary things. It has had its bright period, and its fall, has risen again to a certain degree, and now declines once more. It has not only experienced these alternatives, but has varied even in its fundamental principles; for what was considered at one time as an essential attribute, has been coldly looked upon in a successive age, as of no consequence; besides, there has been a contrariety and difference of opinions in various times, relating to those points which constitute the very art itself. (P, 43-44)

So alongside his principled commitment to canonic art Mengs had a sure sense of the relativity of times: 'every age has its particular character, which by means of a general ferment, enlivens the imaginations of men' (2:137). But the relative and the normative intersect, according to Mengs, or at least cannot gainsay the fact that the Greeks were the acme of art. The Romans ruined art through opulence, and a barbarous taste arose when Christianity was embraced (2:17-19). The arts revived with Giotto and the study of perspective; the quattrocento artistic geniuses in Italy became models of perfection. After 'a pause,' writes Mengs, the Carracci, especially Annibale, reintroduced 'the justness of the Greek statues' to painting (P, 54); Poussin too 'came nearest to the classic style of the Greeks' (P, 57). But Pietro da Cortona, three generations ago, and scholastic disciples of the great painters became superficial, extravagant, and led to the recent ruin of painting (2:130-46; P,55-56).

That this ground plan became standard a generation later need not concern us here.[36] That Mengs shared Winckelmann's belief that climate brooked large in encouraging or limiting artistic performance and that 'political liberty' and the 'customs' of a land served either as a base for a flourishing artistic community or impaired it (2:3-5, 20-21, 138-39, 144, 147) can be documented. But important for understanding the fruitfulness and the latent sterility of Mengs' neoclassical paradigm about the recurrence of canonic art is to ask pointedly with respect to the writing of art history: does the conception that Greek sculpture and painting was the pinnacle of artistic excellence, followed by a fall, and that the quattrocento showed perfection, pause, flurry ... and then a fall, mean that this is an inevitable, recurrent pattern?

Mengs's categorical framework of style-types holds steady, it seems, during changes in painting.[37] Mengs does not convert his rubric of styles into schematic phases of a lockstep sequence as Winckelmann does when the latter traces the changes in ancient art from the Archaic (*ältere*) style rising to the Lofty (*hohe*) style, slipping to the Lovely (*schöne*) style and down to the discredited Copyist style (*der Nachahmer*).[38] And Mengs does seem to allow for historical openings and resurgencies that have a chance nature – 'accidental causes' and 'good fortune' (2:140). But he maintains as axiomatic an underlying art historical principle that sets up a final dilemma: since all human thought and action tend toward progression, once perfect art has been reached by a Master artist every attempt to 'superadd' to what has been achieved – whether out of the competitive desire to surpass, or a genuine attempt to improve – the 'too much excellence' introduces what is unessential, novel, superfluous, and therefore renders the ensuing art faulty (2:117, 122, 130-31, 136-67, 146, 152; P, 49, 115-16).

The dilemma is this: If canonic art is attainable at large in history more than once, decline will be periodic. If the rise and fall of painting is a permanent recurrence (also palimpsestic?), artistic leadership that asks one to learn the Beauty of Nature by imitating the Master artist leads one into a bind (2:106-7). If there is a canon for painting, then there is an entelechial closure to its possible development and epigones are ontologically inescapable. If there is an openendedness to what painting can become, is there then a binding canon?

Mengs's art historiographic account remains troubled by this dilemma. He is unsure about whether Raphael, who was the 'most divine,' entails that Correggio was lesser, or whether Correggio's 'adding' charm to Raphael's work which never reached for the grace beyond Expressive art means Correggio was not still more superlative (1:54-55; 3:52-53, 149; P, 53). Mengs's fundamental commitment in theory to canonic art, however, and his attempt at a millenial artwork in *Parnassus* attest to his desire to make the palimpsestic dilemma fertile. Mengs meant to restore to patrons and the up-and-coming generation of artists the accumulated treasure of art knowledge 'hidden in forgetfulness' (2:19). He accepted this Sisyphean task without demur as a good neoclassical Idealist should, until at age fifty-one he died from overwork.

The dilemmas in Mengs' thought and work bore the fruit of an art historiographic method because he dared attribute relational categories to art which was regarded as normative during cultural change.

That Mengs identified *a certain tradition* (the fictional 'Greek' one represented by Raphael and Correggio) (2:91-92, 101-3) to be *the canon*, and misjudged innovation by a norm of *simplification* were neoclassical errors which impeded his insights. That he did face the dilemma of giving leadership in theoretical aesthetics, artistic style, and art historiography at a time of considerable cultural stress at least makes us aware today that there are such problems.

CALVIN SEERVELD
Institute for Christian Studies, Toronto

Notes

1 Cf Nikolaus Pevsner, *Academies of Art past and present* (Cambridge: Cambridge University Press, 1940), pp. 150-51.

2 J.N. de Azara's generous comment 'que todo lo que hay technico en la Historia del Arte de Vinkelman es de su amigo Mengs' ('Noticias de la vida y obras de Don Antonio Rafael Mengs,' in *Obras de D. Antonio Rafael Mengs*, ed., Joseph Nicholas de Azara [Madrid: Imprenta Real de la Gazeta, 1790], p. xxxv) is based on Mengs' undisputable connoisseurship. Cf Mengs' letter to Fabroni (in *Opere*, ed. Carlo Fea [Roma: Stamperia Paglarini, 1787], pp. 357-68); Otto Harnack, 'Raffael Mengs' Schriften und ihr Einfluss auf Lessing und Goethe' (1892) in *Essais und Studien zur Literaturgeschichte* (Braunschweig: Verlag von Friedrich Vieweg und Sohn, 1899), pp. 200-201; and Herbert von Einem introduction to *Mengs, Briefe an Raimondo Ghelli und Anton Maron* (Göttingen: Vandenhoeck & Ruprecht, 1973), pp. 20-21. Thomas Pelzel (in 'Winckelmann, Mengs and Casanova: a reappraisal of a famous eighteenth-century forgery,' *Art Bulletin* 54 [1972], 300-15) might find support for his judgment on the apocryphal nature of Mengs' role in the incident of fooling Winckelmann in the fact that the original Spanish memoir by de Azara *lacks* the account of the incident which is given in the 1783 Italian Bassano edition on pp. lxxxiv-lxxxix.

3 Cf Goethe's letter of March 1788 as he was reading the new 1878 Fea edition of Mengs's works: 'Es ist in allem Sinne ein trefflich Buch ... Auch seinen Fragmenten über die Schönheit, welche manchem so dunkel scheinen, habe ich glückliche Erleuchtungen zu danken' (in *Italienische Reise 1786-1788* [München: Hirmer Verlag, 1960], pp. 507-08). Also, Goethe's letter of 17 September 1799 to Knebel: 'Ausserdem habe ich jetzt mit Meyern die Kunstgeschichte des gegenwärtigen Jahrhunderts vor. Erst bis auf Mengs und Winckelmann, dann die Epoche die sie machten, und welche Wendung nach ihnen die Sachen genommen habe' (*Briefe*, ed. Mandelkow [Hamburg: Christian Wegner Verlag, 1964] p. 397. Cf Herbert von Einem, 'Goethe und die bildende Kunst' (1949) in *Goethe-Studien* (München: Wilhelm Fink Verlag, 1972), pp. 122-30.

4 For convenience sake the complete English translation of the *Works of Anthony Raphael Mengs*, ed. J.N. de Azara, translator unknown (London: R. Faulder, 1796), is cited in the text by volume: pages.

5 This lengthy letter treatise to Ponz 'sobre el merito de los Quadros mas singulares que se conservan en el Palacio Real de Madrid' (1776) will be cited in the text from the John Talbot Dillon English translation from the Spanish and published as *Sketches on the Art of Painting* (London: R. Baldwin, Pater Noster Row, 1787) as (P, pages).

6 Pelzel depreciates Mengs' 'gospel of eclecticism' (*Anton Raphael Mengs and Neoclassicism; his art, his influence and his reputation* [Princeton University Ph.D. diss. under R. Rosenblum, 1968], pp. 117-19) out of a bias that shows itself later: '... in the final analysis, Mengs can only be considered, in both theory and practice, as having been eclectic and conservative rather than imaginative and revolutionary' (*ibid.*, pp. 348-49). The fact that one man's boon is another man's bane and depends much upon the reader's own subjective perspective is documented by the opposite judgments of Mengs' younger contemporaries Carl Ludwig Fernow (cf *Neuen Teutschen Merkur* of 1795, quoted in Dieter Honisch, *Anton Raphael Mengs und die Bildform des Früh-Klassizismus* [Recklinghausen: Verlag Aurel Bourges, 1965], p. 15) and Goethe (cf supra n. 3).

7 The original source of this listing of elements for painting is probably found in book II of Leon Battista Alberti, *De pictura* (1435), par. 30-31, 35, 46: 'Picturam igitur circumscriptio, compositio et luminum receptio perficiunt.'

8 Cf Ulrich Cristoffel, *Der Schriftliche Nachlass des Anton Raphael Mengs. Ein Beitrag zur Erklärung des Kunstempfindens im spätern 18. Jahrhundert* (Ph.D. diss. under H. Wölfflin [Basel: Benno Schwabe & Co., 1918]), p. 139 n. 18.

9 An example: Lodovico Dolce, *Dialogo della Pittura intitolato l'Aretino*, cited in W. Tatarkiewicz, *History of Aesthetics* (Hague: Mouton, 1974) 3:214: 'Deve il pittore procacciar non solo di imitar ma di superar la natura.' Mengs: '...si vedrà, che l'Arte può molto ben superar la Natura' (*Opere* [Bassano 1783 edition], 1:19).

10 Note (a) 'Stimava più di tutti il Lomazzo. FEA' in *Opere*, ed Fea (1787), p. xxxv.

11 G. Paolo Lomazzo, *Idea del tempio della pittura* (1590), chapter 17: 'Mà dirò bene che à mio parere chi volesse formare due quadri di somma profetione come sarebbe d'vno Adamo, & d'un Eua, che sono corpi nobilissimi al mondo; bisognarebbe che l'Adamo si dasse à Michel Angelo da disegnare, à Titiano da colorare, togliendo la proportione, & conuenienza da Rafaello, & l'Eua si disegnasse da Rafaello, & si colorisse da Antonio da Coreggio: che questi due sarebbero i miglior quadri che fossero mai fatti al mondo' (photographic reproduction, Hildesheim: Georg Olms Verlagsbuchhandlung, 1965, p. 60). Cf Rensselaer W. Lee, *Ut Pictura Poesis: The humanistic theory of Painting* (New York: W.W. Norton & Co., 1967), pp. 9-16.

12 1762, 1765, 1771, 1774, 1778. Cf Winfried Lüdecke, 'Mengs-Bibliographie,' *Repertorium für Kunstwissenschaft* 40 (1917), 255-60.

13 Cf Tatarkiewicz on Lomasso, *History of Aesthetics* 3:208; and Erwin Panofsky on Bellori in *Idea, a concept in art theory* (1924), trans. J.J.S. Peake (Columbia: University of South Carolina Press, 1968), p. 106.

14 Cf J.N. de Azara, 'Noticias' in *Obras* (1780), pp. xxvi-xxviii.

15 '... those pieces which are forced or overcharged, destroy the taste of the art, but the simply beautiful works accustom the eye to a more delicate sense ... The best Taste which Nature can give, is that of the Medium, since it pleases mankind in general' (1:26). Winckelmann articulates the thought two years later in 1764 with choice images: 'Durch die Einheit und Einfalt wird alle Schönheit erhaben ... Diejenige Harmonie welche unsern Geist entzückt, besteht nicht in unendlich gebrochenen, gekettelten und geschleisten Tönen, sondern in einfachen lang anhaltenden Zügen ... Nach diesem Begriff soll die Schönheit sein, wie das vollkommenste Wasser aus dem Schoosse der Quelle geschöpft, welches, je weniger Geschmack es hat, desto gesunder geachtet wird, weil es von allen fremden Theilen geläutert ist' (*Geschichte der Kunst des Altertums* par. I, iv. 2.1 [Leipzig: Verlag der Dürr'schen Buchhandlung, 1882], pp. 110-11).

16 Cf Otto Harnack, p. 199; Wilhelm Waetzoldt, 'Mengs als Kunsthistoriker,' *Zeitschrift für Bildende Kunst* N.F. 30 (1919), 122.

17 Winckelmann had discussed and enjoined this option already in his own *Gedanken über die Nachahmung der Griechischen Wercke in der Mahlerey und Bildhauer-Kunst* (1755): 'Scheinet die Vorstellung möglich, so ist sie es nur allein durch den Weg der Allegorie, durch Bilder, die allgemeine Begriffe bedeuten' (in *Kleine Schriften, Vorreden, Entwürfe*, ed. Walther Rehm [Berlin: Walter de Gruyter & Co., 1968], pp. 55-56). Winckelmann developed the theme again in *Versuch einer Allegorie, besonders für die Kunst* (1766).

18 'Mengs, Alessandro Albani und Winckelmann — Idee und Gestalt des Parnass in der Villa Albani,' *Storia dell'arte* 30/31 (May-December 1977), 87-156 +60 plates.

19 S. Röttgen, pp. 114-21.

20 Cf Carl Justi, 'Der Cardinal Alexander Albani,' *Preussische Jahrbücher* 28 (1871), 338-43.

21 'Der Künstler hat ein Werck vonnöthen, welches aus der gantzen Mythologie, aus den besten Dichtern alter und neuerer Zeiten, aus der geheimen Weltweissheit vieler Völcker, aus den Denckmählern des Altertums auf Steinen, Müntzen und Geräthen diejenige sinnliche Figuren und Bilder enthält, wodurch allgemeine Begriffe dichterisch gebildet worden' (Winckelmann, 'Gedanken' [1755], in *Kleine Schriften*, p. 57). 'Der Pinsel, den der Künstler führet, soll im Verstand getunckt seyn, wie jemand von dem Schreibe-Griffel des Aristoteles gesaget hat: Er soll mehr zu dencken hinterlassen, als was er dem Auge gezeiget, und dieses wird der Künstler erhalten, wenn er seine Gedancken in Allegorien nicht zu verstecken, sondern einzukleiden gelernet hat' (*ibid.*, p. 59). 'Eine jede Idee wird stärker, wenn sie von einer oder mehr Ideen begleitet ist, wie im Vergleichungen, und um so viel stärker, je entfernter das Verhältniss von diesen auf jene ist ... Je mehr unerwartetes man in einem Gemälde entdecket, desto rührender wird es; und beydes erhält es durch die Allegorie' ('Erläuterung der Gedanken von der Nachahmung' [1756], in *Kleine Schriften*, p. 120).

22 Cf S. Röttgen, pp. 140-43.

23 *Ibid.*, p. 128.

24 Cf '... an insipid and artificial style, as may be appreciated from Mengs' own *Parnassus* (Villa Albani, Rome, 1761)' in *The Oxford Companion to Art*, ed. Harold Osborne (Oxford: Clarendon Press, 1970), p. 768. Hugh Honour's insightful and

magisterial summary of 'Classicism and Neo-classicism' does, however, fail on this point too: 'Belonging to the early, negative anti-Rococo phase of Neo-classicism, it [Mengs' *Parnassus*] seeks to do nor more than recreate a dream of classical perfection by a synthesis of antique sculpture and Raphael's paintings' (*Neo-classicism* [Harmondsworth: Penguin books, 1968], p. 32).

25 Röttgen is correct too in distinguishing the 'borrowing' of a motif or two, which Gerstenberg detects (*Johann Joachim Winckelmann und Anton Raphael Mengs* [Halle: Max Niemeyer Verlag, 1929], p. 14), from the 'completely other, over-all cast of the determining style' of what Mengs produced in contrast to the discovered pieces Mengs came to see. Cf S. Röttgen, p. 140.

26 Röttgen grounds and motivates historically the rigorous exceptionality of the *Parnassus* fresco and attributes its uniqueness to Winckelmann's dictation on this particular commission; cf *Storia dell'arte* 30 (1977), 137-39. Dieter Honisch downplays its Sonderstellung most, perhaps, in his formcritical problematics that stresses the continuity of Mengs' shift from ornamental motifs to use of the painting field for ordering the composition (cf *A.R. Mengs und die Bildform*, pp. 34-37). Pelzel debunks Mengs' whole contribution to 'the reformatory classicism' as only an 'episodic,' 'incidental factor' in Mengs' artistic corpus (*A.R. Mengs and Neoclassicism*, pp. 127, 367).

27 'The British contribution to the Neo-classical style in painting,' *Proceedings of the British Academy* 40 (1954), 58-59.

28 Cf Honisch, pp. 22, 26, although Honisch calls Mengs' Dresden altarpiece on 'Christ's Ascension' (1755-66) and the San Eusebio fresco (1757-58) loosely 'noch in barocker Tradition,' p. 27.

29 Another painting of these years reinforces the judgment; cf Pelzel on 'Augustus and Cleopatra,' pp. 99-108.

30 Honisch notes how Mengs appropriated Correggio's themes and even formal properties in his 'Passion scenes' of 1765-68 (pp. 31-32). Pelzel sees the Correggian hegemony in later Mengs as evidence for denying Mengs any serious role in the spread of neoclassical art (pp. 145-52, 348-52). But Pelzel finds the Parnassus in fact already retrogressive toward a Rococo formal *Empfindung* (pp. 126, 140)! and reads the late Mengs loosely back into 'the Baroque tradition' (p. 165). Herbert von Einem shows how Mengs struggled to modify neoclassically, as best he could, the heroicizing project of decorating the Spanish King's palace, e.g., the heroes in 'Hercules' Apotheosis' (c. 1762-75) sit, stand, and walk around, separate from the clouds, and weightless Doric columns replace the heavens in the 'Apotheosis of Emperor Trajan' (c. 1774-76) (*Mengs, Briefe an Raimondo & Maron*, pp. 17-18). Every art historian stumbles to find accurate terms: Gerstenberg uses 'baroque classicism' (p. 27), and S. Röttgen 'Neo-seicentismus' (p. 145).

31 Rosenblum opens the door to this insight by identifying 'the Neoclassic Horrific,' 'the Neoclassic Erotic,' and 'the Neoclassic Stoic' – all modes of Neoclassicism; but he skews this insight when he understands at least 'the Neoclassic Horrific' to have a *Romantic* view of antiquity (cf *Transformations in late eighteenth century art* [Princeton: Princeton University Press, 1967], pp. 11, 20, 28). 'Neoclassic' and 'Romantic' were contemporary trends or 'periods,' in my judgment, and both looked back in time; but their driving forces excluded each other, and were as at odds as Sir Joshua Reynolds and William Blake. Cf C. Seerveld, 'Towards a car-

tographic methodology for art historiography,' *Journal of Aesthetics and Art Criticism* 39 (no. 2, 1980), 148-49.

32 Pelzel translated *estilo viziosos* as 'Vicious Style' (p. 144). This slanted translation exemplifies the tendentious way Pelzel twists so many facts about Mengs's thought, making Mengs seem mean-tempered or simple-minded. Mengs is not, as Pelzel claims, 'most contemptible of all' on 'the Easy Style' either. Mengs's serious effort to give good painterly advice and tips on taste is very undoctrinaire and generous. On the 'vitiated style' Mengs writes: '... I do not condemn the authors, since great defects are often annexed to great accomplishment ...' (P, 15).

33 J.N. de Azara, 'Noticias,' p. xxxix.

34 Herbert von Einem seems to interpret the types of style as a psychological index of artistic dispositions: 'zwar in psychologischer Hinsicht als Unterscheidung von Begabungstypen' (*Mengs, Briefe an Raimondo & Maron,* p. 18). While Mengs gives cause for such a reading occasionally ('I advise therefore all painters not to study Correggio if they have not sensibility like him.' 1:67), Waetzoldt sees the issue more sharply when he refers to the delineated Urstile as 'Anschauungs- und Gestaltungsweise' (*Deutsche Kunsthistoriker* [Leipzig: E.A. Seemann, 1921], 1:91-92). There may be the ready-made anthropology of *facultates* and the still earlier 'humours' psychology of Hippocrates, which was so popular in the Renaissance, behind Mengs' typology; but Mengs' schema is intent upon *artistic* matters more than on a roster of personality-archetypes.

35 Reynolds's *Discourses* come later (from 1769 on) and are a comparable, if less systematic, attempt to give structure to 'learning' critique and the history of art.

36 Cf Alexander Potts, *Winckelmann's Interpretation of the History of Ancient Art in its Eighteenth Century Context* (Warburg Institute, University of London, Ph.D. diss. under E.H. Gombrich, 1978), 2:476-78.

37 In fact, Mengs uses the abiding *Hauptgattungen des Stils* or *Sehweisen* to detail the change in a given painter's style: Velasquez shifts from painting in the Natural style to a Graceful style, to the Expressive and finally to the Beautiful style (2:82-83; P, 62-63); Murillo turned from the Natural style to one of sweet Loveliness (2:84; P, 64-65); Raphael went through three styles, the last and best was developed out of Raphael's appreciation for Michelangelo's Sistine chapel painting (2:133); in Correggio too Mengs carefully discerned three changes of style (3:26, 38).

38 *Geschichte der Kunst des Altertums* par. I, iv, 3. 1-4, pp. 152-73. Cf Carl Justi, *Winckelmann und sine Zeitgenossen* (Leipzig: Verlag von F.C.W. Vogel, 1923) 3:154-58.

9. Musical Rhetoric and Other Symbols of Communication in Bach's Organ Music[1]

This paper illustrates certain communicative devices in Bach's organ music. Such devices are usually designated by the term 'symbol.' This term is employed in the present paper in the broadest sense; it denotes a means to communicate something – directly and indirectly. This proviso is necessary since the word is most frequently used in a restricted sense to indicate an indirect form of musical communication, i.e., a musical sign which, representing something beyond itself, requires familiarity and abstract decoding before it is understood. In itself such a sign would be more or less neutral and the idea, which it represents, understood in the abstract, rather than experienced. There are, however, musical signs (symbols in a wider sense) which communicate ideas that do not require decoding. Concepts such as 'high,' 'low,' 'fast,' 'slow,' 'fear,' 'joy,' 'contrast,' 'repetition' and 'climax' have been experimentally proven to be directly visualized or experienced through musical sound without the need for any additional decoding or translation in the abstract.

It will be demonstrated that Bach's organ music makes use of both forms of communication: the direct manner employed in musical rhetoric, as well as the indirect process of rather esoteric symbols. Musical rhetoric will be illustrated by means of two of Bach's 'free' (textless) organ works, the other symbols by means of the textual chorale preludes (hymn arrangements). The chorale preludes will demonstrate that Bach usually transmits an idea in both ways simultaneously: through rhetoric and the creation of moods, as well as through the hid-

MAN AND NATURE / L'HOMME ET LA NATURE III/1984
0824-3298/84/0300-0131 $01.25/©C.S.E.C.S./S.C.E.D.S.

den symbols of number, allegory, etc.. The paper will not attempt to prove the same duality for the free, textless organ music (Preludes, Fugues, Toccatas, etc.). Attempts have been made to discover various hidden symbols in Bach's keyboard Preludes and Fugues (semantics of number, number cabalistics, structural symbolism), but, although interesting statistics have emerged, the results remain somewhat speculative in the absence of any contextual reference (text association, liturgical context, etc.). The exception may be the *'St. Anne's'* organ Prelude and Fugue (BWV 552) which act respectively as introduction to and conclusion of a set of liturgical chorale preludes (cf. below).

Essential for an understanding of the embodiment of symbols in Bach's organ music is a historical perspective. A brief survey will be given of the musical tradition and musical concepts from which such a practice sprung.

Finally, the paper does not intend to explain Bach's music completely in terms of rhetoric and other symbols or to suggest that symbols are present or equally prominent in each work. His music is, foremost, music to be enjoyed and analyzed in musical terms: theory, harmony, counterpoint, style, ornamentation and musical form. Rather the paper illustrates the extent to which these musical means were influenced by various non-musical symbolic concepts and that the idea of 'absolute music' or 'pure music' needs some qualification as far as Bach is concerned. With few exceptions his music was not written for aesthetic contemplation solely, but to fulfil a specific function. His sacred music served a didactic purpose within the Lutheran liturgical tradition. Although some of his later music moved in a more abstract direction, Bach did not subscribe to the notion of 'autonomous' or 'absolute' music which was later developed in the Classic era. Those who prefer to interpret Bach's music as absolute and non-referential are, nevertheless, correct in one regard: even when he applied rhetorical concepts within the traditional learned music style, the tonal means and sound impressions of his music remain intrinsically musical and musically satisfying.

I. Musical rhetoric in Bach's organ music

Since rhetoric is no longer practised widely or taught in our schools as a formal subject, a brief exposition of this art and its general application to music might clarify the subject matter.

Rhetoric is 'the art of using language so as to persuade or influence others; the body of rules to be observed by a speaker or writer in order that he may express himself with eloquence' (OED). It must be based on logic and dialectics yet it must also appeal to the emotions and passions of the hearers. The power of persuasion has undoubtedly been realized and intuitively practiced since the beginning of human history; it was formally recognized in the earliest Greek writings, it developed into a formal art during the Greek classical period, and it was refined by Aristotle, Cicero, Quintilian and others.

This art comprises four stages (Note: The following rhetorical terminology is that used in Germany during the times of Bach and deviates somewhat from Quintilian's and other writers' of Antiquity):

1) the collecting or finding of facts or material (Inventio); several resources, guidelines ('loci topici') exist for this purpose
2) the orderly arrangement, disposition of the material collected (Dispositio, Elaboratio): Exordium (introduction), Narratio (narration), Propositio (statement of view), Confirmatio and Confutatio (confirmation of view and refutation of opposite views by means of analysis), Peroratio (conclusion)
3) the use of the right words and emotional decoration or presentation of the facts (Decoratio, Ornatus), the creation of moods by means of more than 160 different tropes and figures, e.g., repetition, exclamation, rhetorical question, hyperbole, antithesis, emphasis, etc.
4) the memorizing and delivery of speech, the manner of presentation (Memoria; Elocutio, Pronuntiatio, Actio), by means of the proper articulation, emphasis, gestures and emotional involvement[2]

Music and rhetoric inherently have much in common. They

1) need time to be presented
2) consist of continual changes in their sound elements
3) have a beginning, logical continuation and peroration
4) employ rhythm, tempo, the pause, dynamics, repetition and antithesis
5) make use of the law of 'extending parts': statement, statement, extended statement
6) use analogical formulas, e.g., the comma ('imperfect' = half-cadence or incomplete cadence in music), semicolon (fermata), period ('perfect' cadence, complete cadence), exclamation mark (deceptive cadence, interrupted chord)
7) need a reciter or orator (recitalist, performer)

8) make use of moods and emotions

9) need articulation, phrasing and emphasis[3]

Translated into musical terms the four stages of rhetoric as they appeared during the times of Bach were:

1) the finding of a musical idea or motif. In textual music the words can serve as a guide with regard to the spirit of the work and the nature of the motifs and themes to be used.[4] Several 'loci topici' (literally 'places' to find material for a musical 'topic') were suggested by Mattheson, a contemporary of Bach.[5]

2) the arrangement or disposition of the musical ideas.
Mattheson again quoted the arrangement of rhetoric as model:[6]

Exordium	Introduction (to main theme, e.g., by motifs)
Narratio	Narration (where applicable)
Propositio	Statement of main idea or theme
Confirmatio, Confutatio	Development and analysis of the main theme by means of motif-fractionation; refutation of new and contrary themes, motifs, keys, dissonants; confirmation of main idea and key
Peroratio	Conclusion

This disposition proves to be very similar to the later Sonata-form-idea so often used in the first movement of the Classic sonata and symphony.

3) Decoratio, the emotional presentation of the ideas, the creation of moods by means of music.[7] Apart from certain purely musical means like key-relationships,[8] intervals,[9] rhythms,[10] tempi,[11] and embellishments,[12] the different tropes and figures were translated into music where possible, thus creating approximately eighty-two different musical tropes and figures (subdivided into grammatical, pictorial and affective figures)[13] — see Addendum I.

4) Elocutio, the presentation or recital of music itself. Like rhetoric, emotional involvement is required in music, also the right amount of emphasis, clarity of articulation and meaningful gestures. The latter are to a great extent composed in the music, but the performer has to give expression to these without affectation.[14]

The practice of music as a rhetorical art during the time of Bach drew its essential impulse from the still predominantly theocentric philosophy of the early eighteenth century. Music served to instruct the listener (religious instruction in the case of church music, as will be illustrated later),[15] but relied on a long tradition of musical persuasion.

The psychological and therapeutic use of music was known and practiced in the East and Asia Minor during the second millenium B.C. (e.g., David's music for Saul) and further cultivated in Greece by Pythagoras and others. The Greek *Musike* did not denote pure music but a complete integration of word, musical sound and dance.[16] Its influence was such that it was regarded as a powerful ethical force. Plato gave final shape to this concept in his so-called Ethos-doctrine. Certain music and musical instruments were regarded as morally uplifting and were recommended for pedagogical purposes; others, associated with intoxication, orgies and eroticism, were regarded as decadent and prohibited in schools. This Ethos-doctrine was based on the concept of mood (*affectus*), on Hippocrates' doctrine of the four temperaments and on the prevailing speculative cosmic theories (see below pages 152-54). It was upheld during the medieval period (although music no longer formed an integral part of the dance but showed an increasing independence from the text as it developed polyphonically), but translated into Christian terms. The Church encouraged sacred music and discouraged secular music, especially the rhythmic dances of the jugglers. Renaissance humanism, with its interest in and imitation of nature and human nature (*imitatio naturae* included the physical and psychic qualities of man), contributed to a renewed study of the influence of music on human nature, but the emphasis shifted: the musical Ethos-doctrine was transformed into the sixteenth century doctrine of moods (Glarean, 1547; Zarlino, 1558) which dominated the musical thinking of the next two centuries. It was a change in emphasis, from the ethical to the more psychological influence of music.[17] Coupled to this was the Renaissance emphasis on textual clarity, an endeavour to integrate music and words. Textual music of the Musica Reservata (sixteenth-century polyphonic music following Josquin Desprez) endeavoured to match the syntax of the words and simultaneously capture its mood, becoming more rhetorical in essence.[18]

The spiritual climate of the Counter-Reformation, as embodied in the Council of Trent (1545-1563) and the creation of the Order of the Jesuits, further contributed to a rhetorical emphasis in music — at a time when Italy became the new centre of musical development. With its renewed emphasis on the transcendental and on the sinfulness of human nature, the Church actively promoted art which stirred the human conscience and focused on an eternal destiny. The arts became didactic in purpose. This dramatically strengthened the already rhetorical element in sixteenth century music. Instead of the Renaissance ideal of balance between text and music, the text dominated music during the Baroque era (ca. 1560-1740); the emotional restraint and the moderate moods displayed in the music of the Renaissance gave way to the most extreme, contrasting moods — hate, love, exultation, depression, etc. — which shocked the listener into attentiveness.[19] (This change in the concept and purpose of art was later viewed as a decadent phase of the Renaissance and labelled as 'exaggerated,' 'bizarre,' 'extreme,' 'baroque' by the philosophers of the eighteenth-century Enlightenment.)[20] The foundations for a rhetorical concept of music were completed with the development of the monody (accompanied solo song by means of a harmonized, figured bass or *basso continuo)* around 1600, the accompanied musical recitative (which is an imitation in music of natural speech),[21] and the new antiphonal style of music of the Gabrielis and the *concerto grosso* (which created dialogues between vocal or instrumental groups). In Italy rhetoric was applied more intuitively than in Germany where theorists had been discussing and documenting its musical application since the late 1500's. Music became in theory a *musica poetica,*[22] a discipline of rhetoric rather than of mathematics (as it had been in the Ancient and Medieval period, as will be later illustrated). A diagram may illustrate the relation of subjects within the German school system in which Bach was instructed in Lüneburg (see Addendum II): theology forms the pinnacle of this educational pyramid, the subject which gives perspective to all study. Micro- and macrocosmos are seen as the creation of God, the purpose of existence being the glorification of God. All subjects acquire a meaningful relation within God's order. Music becomes instrumental as a discipline of rhetoric to instruct, to communicate, to convert.[23]

With rhetoric as an important approach to musical composition, composers continued and intensified the late-Renaissance trend of creating moods through increasing translation of various tropes and figures of speech into music. Even the structure of a speech (Dispositio) was adopted, leading to the concept of *Klang-Rede* ('musical speech') in Bach's time with regard to instrumental music.[24] The various theoretical treatises from the late sixteenth to the middle of the eighteenth century demonstrate an increasing dominance of rhetoric over music.[25] Analysis of Baroque music itself confirms this trend. This dominance was broken only by eighteenth-century naturalistic philosophy, which seriously challenged a strictly Biblical philosophy and rejected the complex symbolism and transcendentalism of Baroque art (see also Scheibe's critique of Bach, later in this paper).[26] This resulted in the new classical concept of an 'autonomous music' (Goethe),[27] music no longer dominated by the text or other extra-musical influences (e.g., Mozart's comments on music, especially the opera).[28] A rhetorical concept of music nevertheless lingered on through the early Classics and Beethoven.

Bach's own world view was rooted in the orthodox theology of the Lutheran Church, even if it was a view influenced by pietism and the new currents of rationalism, thus showing discrepancies of which he was unconscious (as Blankenburg and Blume have illustrated).[29] His Christian philosophy is revealed in various ways. In the tradition of Luther, music always had to be functional. Whether religious or secular it appeared under the heading 'Soli Deo Gloria.'[30] Music thus dedicated to the glory of God did not know the boundaries between sacred and secular; only the text put it into one of these categories. Perhaps for that reason Bach adopted many of his secular cantatas as sacred cantatas and vice versa without regarding this as sacrilege. He wrote in his 'Gründlicher Unterricht des Generalbasses,' 1738: 'Like all music the thorough-bass should have no other aim, end or motivation, than to be to the glory of God and the recreation of the human spirit. If this is not taken into consideration, we do not have real music but an infernal noise.'[31] This Christian world view remains uncompromised by Bach's seemingly contradictory actions, e.g., his ill-temper, his personal clashes with his employers, even his resentment against his unenviable church position in Leipzig, for it transcends the sphere of human weakness.[32] What Bach meant by 'the recreation of the human spirit' is spelt out clearly in the foreword to his *Orgelbüchlein*: 'Dem höchsten Gott allein zu Ehren, dem Nächsten, draus sich zu belehren.' It is significant that he used the biblical words 'Nächsten' (fellowman) and 'belehren' ([self]-'instruct' by means of the music). The didactic nature of his music, embodied in rhetoric and symbolism, is a direct expression of his Christian philosophy.

The rhetorical quality of Bach's music is not only to be verified by musical analysis but supported by the testimony of his contemporaries. Johann Nikolaus Forkel, the composer's first biographer (1802), who received first-hand information about Bach and his views from Bach's sons, described the composer as 'the greatest musical poet and the greatest musical speaker in musical history.'[33] He reported that Bach regarded music as a complete 'language,' the individual 'voices' (the instrumental or vocal parts in a composition) as 'persons conversing with one another,' listening to one another and trying to contribute to the 'conversation' in a meaningful way. Apart from his training in musical rhetoric at school, Bach became in later years a personal friend of Johann Abraham Birnbaum, professor in Rhetoric at the University of Leipzig. Birnbaum, in mentioning their various discussions on the relationship between the two arts, wrote: 'Bach was so knowledgeable in the sphere of the similarities which exist between the two arts, as well as in the similar mode and method of composition, that one not only enjoyed listening to him when he pointed this out in a discussion, but one can only admire the clever application thereof in his compositions.'[34] Equally interesting is the account of one of Bach's pupuls, J.G. Ziegler, in 1746, that Bach instructed him not to play the 'Lieder' so superficially but according to the 'mood of the text.'[35] The affective, persuasive character of much of his music, whether religious or secular, vocal or instrumental, can be verified through analyses. His *St. John's* and *St. Matthew Passion* display this rhetorical, didactic trend throughout, confronting the listener with the religious implications of the events by means of subjective application.

In this regard Bach could draw on a rich heritage of musical composition based on rhetoric, from the Musica Reservata, Monteverdi, Schütz to Buxtehude, Kuhnau and others. This tradition, which is also described in the musical treatises mentioned, is an important aid in the analysis of Bach's music. The mood is generally detectable throughout the text in the tropes and figures, key relationship, metre, rhythms, tempo, intervals and dynamics associated with different moods. In textual works the aspect of 'tone painting' and other descriptive means is of importance. The fact that we deal with relatively *objective* criteria for analysis of music in Bach's time is also due to the sociological framework within which this music was written.[36] The composer did not write on the spur of the moment, neither did he display his own changing emotions, but was required (by the court or church) to write music for a specific occasion. The mood was clearly defined by the occasion or the text (something objective with which the composer had to empathize) and he could draw on a general musical vocabulary to achieve this. This is not to suggest that Bach wrote music in a textbook fashion, neither that his music can ever be fully explained and its subtleties captured and described by any analyses.

Two examples from Bach's 'free' (non-textual) organ works might serve to illustrate the rhetorical character of his music: the *G Minor Fantasia* BWV 542 and the *'Dorian Toccata'* BWV 538.[37]

The *G Minor Fantasia* shows a certain affinity to the *St. John's Passion* which was composed around the same time. It displays a similar exultation mixed with deep sorrow expressed by the opening chorus of the Passion. It starts with similar repeating exclamations (Exclamatio, Anaphora, Ex. 1), which are converted into an instrumental recitative (a free musical narrative supported by occasional chords; measure 1 ff., Ex. 2).

Ex. 1 St. John's Passion G Minor Fantasia

Ex. 2

Melodic and harmonic dissonances are employed (Saltus duriusculus, Parrhesia). This is followed by a sequence of repeated and extended motifs (Ex. 3, m. 4-7), an accumulation of ideas and a concluding Seufzer (a descending two-note motif expressing 'sighing,' m. 8-9, Ex. 4).

Ex. 3

Ex. 4

This epic recitative is followed by a contrasting, introverted section in linear counterpoint (horizontal lines of music instead of vertical chords) as a somber 'comment.' The already described 'art of discussion between voice parts' (A, B, C; on a melodiously dissonant bass, D) is cleverly employed (Ex. 5, m. 9^2 - 14^1). The same motifs are used or extended in a discursive way by the various voice parts, and a wealth of figures such as Anaphora, Aposiopesis, Antitheton, Metalepsis, Passus- and Saltus duriusculus is displayed. It ends with a statement (a descending two-note 'sighing motif') similar to measures 8-9 (Epistrophe, Ex. 6, m. 13-14).

Ex. 5

Ex. 6

This contrapuntal section is followed by an agitated continuation of the harmonic recitative with a further display of exclamations, dissonant chords, extension and repetition of ideas (Ex. 7, 8), deceptive cadences and climaxes (e.g., Ex. 8, 9). The figures Exclamatio, Parrhesia, Saltus duriusculus, Anaphora, Paronomasia, Ellipsis, and Gradatio predominate.

Ex. 7

Ex. 8

Ex. 9

Another contrasting 'discussion' section in counterpoint follows (m. 25-31). After this the highly dramatic recitative is continued, employing the same affective figures as before (e.g., m. 35ff: (1) Gradatio, (ii) Pleonasmus, (iii) Ellipsis, (iv) Congeries, (v) Pathopoiia/Parrhesia, (vi) Suspensio, Dubitatio, Ex. 10).

Ex. 10

It is only in the final affirmative major chord that the intense, dramatic tension is resolved.

In a completely different vein, rhetoric is used in Bach's *'Dorian Toccata.'* This piece employs more abstract, eloquent musical motifs than the highly dramatic, affective ones of the Fantasia just discussed. It is a particularly interesting work: an oratory in dialogue form. The musical 'debate' is brought about by a continuous change between the two manuals, which were minutely indicated by Bach. Organ dialogues were very popular even before Bach, but this dialogue is unique in its logical and dialectical lay-out; it uses both the structure of speech (Dispositio) and a wealth of rhetorical figures. The usual rhetorical arrangement (introduction, statement of main idea, refutation of contrary ideas, analysis and confirmation of main idea, peroration) is followed, but clad in dialogue-style. Instead of the main speaker (A) quoting and refuting opposite views, these views are presented by a second speaker (B), represented through the contrasting registration of the second organ manual.

A starts off by developing the main thesis from a simple motif of alternating notes, x (Anaphora, Paronomasia, Ex. 11).

Ex. 11

In bar 5 this statement appears in the bass and is reinforced by emphatic chordal accompaniment (y) in the upper voices (Emphasis, m. 5-7, Ex. 12).

Ex. 12

In bar 7 the now emphasized argument is extended (y+) with the addition of a pre-beat chord and the Propositio concluded with a firm cadence in measure 13 (Ex. 13).

Ex. 13

B starts out by quoting the emphasized premise and then inverting it (turning it upside-down with the emphatic chordal motif, y+, in the bass) e.g., m. 15, 16 (Ex. 14). Such double counterpoint was seen as an important means of creating antithesis or Antitheton by Bach's contemporary, Mattheson.

Ex. 14

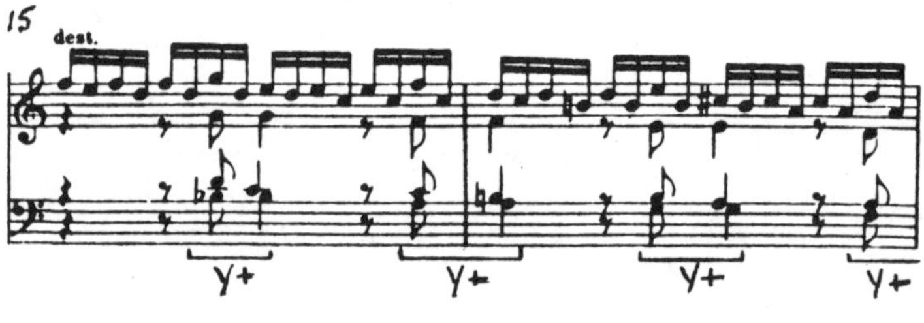

B proceeds by quoting the concluding remarks of A (see m. 10^4 - 13^3, Ex. 13) but extending them (adding another preceding chord to y), which has the effect of exaggeration, generalization (Paronomasia, Mimesis, Epistrophe, Ex. 15, m. 18^2 - 20^1).

Ex. 15

A's reply consists of a parodic quotation and echoing of this exaggeration (y++) alternating between the right and left hand (Anaphora, Mimesis, Dialogismus, m. 20^1ff. – see Ex. 16).

Ex. 16

B now picks up A's introductory 'words' (m. 1-4) but in inverted form (inverting the two lines; Antitheton, m. 25-29, Ex. 17).

Ex. 17

A reaffirms the main thesis by variation and extension (Anaphora, Variatio, Paronomasia), despite opposing interjections from the other who turns the emphatic motif upside down (the ascending motif, A, is changed into a descending one; Mimesis, Antitheton, Exclamatio, m. 29-35, Ex. 18).

Ex. 18

A continues and concludes in a firm cadence (m. 34-37^{1}, Ex. 19).

Ex. 19

B now quotes a small subsidiary section of the last part (see motif q, m. 34-35, Ex. 19). This incomplete quotation (Anaphora, bars 37ff.) is immediately countered by A on a lower pitch (Antitheton, Ex. 20). B nevertheless persists and ends with exactly the same emphatic cadence (m. 43, Epistrophe) as the one A used (m. 36, Ex. 21).

Ex. 20

Ex. 21

A proceeds to extend the initial proposition (adding three preceding notes to the alternating motif x of m. 1 ff.; m. 43, Ex. 22) despite an interjection by B who repeats it on a lower pitch (Paronomasia, Mimesis, m. 43-47, Ex. 23).

Ex. 22

Ex. 23

A now strengthens the main proposition by analysis (fractionation of the main idea into smaller motifs of nine notes, then four notes – Distributio, Ex. 24), variation and emphasis (Variatio, Emphasis), m. 47-66.

Ex. 24

When B's former argument, q, is quoted (cf. Ex. 20), B interrupts A and maintains this viewpoint, ending with the same words (cadence) as before (Antitheton, Emphasis, Epistrophe, m. 66-73).

A repeats the extended thesis, but is directly confronted by B with repetitions of these motifs on a different pitch (m. 76^3 ff., Ex. 25) and then even interrupted before completion (m. 78). At this point A and B speak simultaneously, repeating their 'words.' This heated debate leads to an accumulation of the same viewpoints (Congeries, m. 78 ff.) in an ascending sequence, leading to the climax (Gradatio, Ex. 26), in which A simply 'talks B down' (Pleonasmus, m. 81, Ex. 27).

Ex. 25

Ex. 26

Ex. 27

A follows this up by a spirited confirmation of the main thesis with new extensions and analyses (fractionation of motifs), reaching a victorious climax (m. 90 ff., Ex. 28) symbolized by a change from D minor to D major (m. 94³), the key associated by Mattheson with victory, jubilation and aggressiveness.

Ex. 28

Assured of success, A concludes with strong, emphatic repetitions of the initial alternating motif, x, extended by an ascending, leaping interval (Congeries, Emphasis, Ex. 29).

Ex. 29

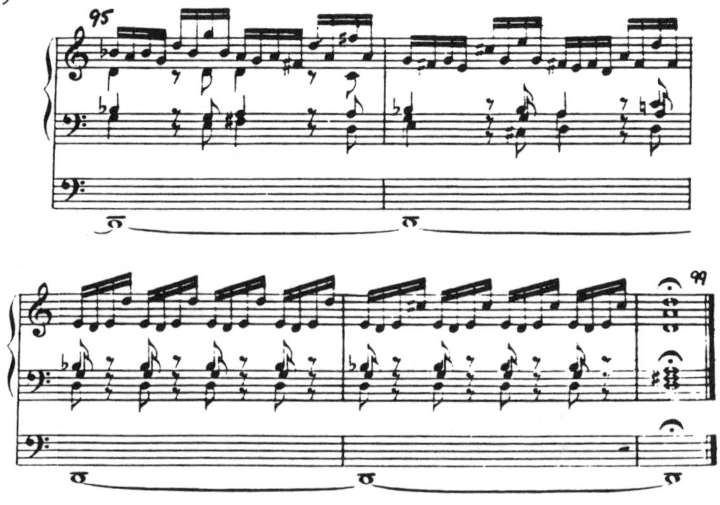

Bach's *'Dorian Toccata'* and the *G minor Fantasia* are only two illustrations of the idea of a purely musical, instrumental *Klang-Rede* in the late-Baroque. As Mattheson wrote (1722): 'Although instrumental works do not employ actual words, nevertheless even the most free and independent works ('Concerten') should contain an expression, an intelligent content, so that they always have something to say and also speak words.'[38]

The idea, sometimes encountered, that Bach's organ music is essentially unemotional, abstract, 'pure music,' music based on 'pure musical form and development' to be played without affect, fails to recognize the philosophical and compositional concepts from which these works originated. But it must be kept in mind that the emotion called for in these works differs from the Storm and Stress of Romantic music. It calls for careful analysis and recognition of the inherent mood of the work. This mood does not change continually as in Romantic music, but remains (apart from subtle shadings) constant for a prelude, fantasia, fugue, choral prelude unless a change in style, tempo, dynamics, etc., occurs. An affective involvement of the player is required to project the mood to the listener.

II. Other symbols in Bach's Organ Music

There are other communication symbols in Bach's music which transcend the dialectical, logical, ethical and psychic levels of rhetoric. They occur especially in his textual organ music, i.e., chorale preludes (hymn arrangements). The employment of these symbols by Bach is equally rooted in his Christian world view and can equally be traced back to concepts dating from the pre-Christian era. They range from fairly direct symbols also employed in rhetoric, such as allegory and imitation, to the more abstract or esoteric ones of emblems, semantics of number and number cabalistics.[39]

1. Allegory, comprising an association of extra-musical contents with musical signs, can be understood by the hearer mainly from the 'sensuous' effect: 'Fall,' 'plunge,' but also 'abyss,' 'sin,' 'damnation' by descending voices, stepwise or by leaps; 'light' and 'dark,' 'heaven' and 'hell' by contrasting high and low registers; 'length,' 'effort,' 'hardship' by sustained rhythms in contrast to the quick rhythms for 'hurry,' 'flying,' 'rapture.' Descriptive allegory can also be classified under the rhetorical 'hypotyposis'-figure and other figures like Antitheton,

Passus- and Saltus duriusculus, Ellipsis, Anabasis, Katabasis, Circulatio.

2. Imitation is the direct copying in music of sounds and noises (chime of bells, birdsong, storm, knocking, etc.).
3. Emblematical, visual symbols are musical signs which translate word-content into tonal figures in a manner which cannot simply be understood from the sensuous effect of those tonal figures, but must be consciously 'known' (or 'seen' in notation); for example: the countless use of the raised tone sign (sharp) in Bach's *Kreuzstabkantata* (the Christian 'bearing of the cross'), as the German word for sharp is 'Kreuz' (in English 'cross' and 'sharp' would never be associated); the use of the note G (or Sol in medieval times) for the Latin word 'sol' ('sun'); 'nox' ('night'), represented by blackened notes, etc.
4. Semantics of number consists in the translation of ideas into musical signs by means of numerical symbols, e.g., seven for Creator and creation, beginning and end; twelve for Church, disciples, congregation; three for Trinity, perfection, the transcendental, etc.
5. Number cabalistics consists of the translation of alphabetical letters into numbers, thus allowing composers to write words or their own names in music by means of the number of notes used in a theme, motif, or entire work or the number of bars used (e.g., Bach = 14, J.S. Bach = 41).

This tradition of visual, mathematical 'musical symbolism' can be traced back to Antiquity.[40] In the ancient Chinese, Babylonian and Egyptian cultures a cosmology of music was developed which was rooted in religious concepts and in the symbolism of number. Musical intervals were expressed in terms of numbers, e.g., the octave being 1:2, the fifth 2:3, the fourth 3:4, etc. These numbers were derived not only from the division of an open lute string in the ratio 1:2, 2:3, 3:4, but they also correspond with the numbers of the natural harmonics ('upper partial tones') obtained by 'overblowing' a wind instrument. In a more speculative way distances between celestial bodies and the relation between the human soul and body were linked together, laying the foundation for a cosmology of music based on number. Consequently, certain intervals influence the human soul, but so do celestial bodies. Stars and planets, moving in a harmonious geometric relation to one another, also create a celestial harmony which is inaudible to the restricted human ear. This concept of a *musica mundana* (celestial harmony) was further developed by Pythagoras and the Greek theoreticians as a counterpart to the equally speculative *musica humana* (the 'harmony between soul and body') and the *musica instrumentalis* (audible music, 'music' in the sense we use the term). The audible music thus became a microcosmic image of a divine creation, a link between this visible world and the transcendental, powerful as an ethical force to uplift or corrupt man by means of the concept of *musica humana*. This speculative, mathematical basis of music became the rather vague motivation for Plato's Ethos-doctrine. Numbers acquired a magic, symbolic meaning not to be tampered with. Pythagoras tried to calculate all intervals by means of the numbers one to four, since $1 + 2 + 3 + 4 = 10$, and ten is a perfect number. This might also explain the initial use of only four strings on the Greek lyre and of four-note scales (*tetrachords*) as basic units. An increase in the number of strings on a musical instrument was regarded as sacrilege. This symbolism of numbers existed also in the ancient Jewish culture and religion; the idea of the creation as something harmonious which sings and praises the Lord partly accounts for the wealth of symbolism in the Old Testament.

With the Christian era these ancient speculative concepts were adopted and explained in Christian terms in the Alexandrian theology. As in Greece and Rome, music remained a discipline of mathematics in the medieval system of the seven *Artes liberales* (the mathematical sciences constituting the *Quadrivium* being Geometry, Arithmetic, Astronomy and Music; the linguistic subjects constituting the *Trivium* being Grammar, Dialectics and Rhetoric). In Biblical symbolism the number three is seen as perfect; with the introduction of rhythmic modes in the twelfth century only triple metre and threefold subdivisions of note values were used. When binary rhythm is introduced in the fourteenth century, it is seen as part of the secular tendencies of the *Ars Nova* and is censured by the church, since the number two is 'imperfect,' hence of a 'wordly' nature. The use of semantics and other forms of musical symbolism is retained in the Renaissance in the form of riddle canons and other manneristic devices (e.g., Ockeghem), and is combined with sound imitation (the new rationalistic *imitatio naturae*) and the creation of moods (especially following Josquin in the early sixteenth century).[41]

When music became a discipline of rhetoric (part of the linguistic sciences) during the Baroque, the art of musical symbolism, semantics, imitation and a cabalistics of number did not die out; it received new stimulus from the renewed emphasis in the arts on the metaphysical and the transcendental. The seventeenth century theoreticians (such as Mersenne[42] and Kircher[43]), therefore, still conveyed the ancient speculative concept of a musical cosmology (*Musica mundana*), of a celestial, inaudible harmony based on number. Similarly, composers and theoreticians display a profound knowledge of the traditional symbolic arts. In the introductions to his compositions and an unpublished treatise Bach stressed the functionalism of music in terms of a deeper symbolism. According to Blume, music that reflects in its number and rule the divine world order, that is based on the *harmonia aeterna* and laws of nature, is for Bach real functional music which deserves to exist. In his 'Instruction in Figured Bass' Bach describes musical harmony built on the Generalbass in terms of such religious symbolism (see above). Studies of various treatises by contemporaries of Bach like Werckmeister[44] have revealed a musical practice still actively employing musical symbolism. Analyses of Bach's textual music, and more recently of his purely instrumental compositions, have indicated a wealth of latent symbolism, semantics and cabalistics of number, allegory and imitation.

In his *Clavierübung*, 3rd Part, consisting of various organ choral preludes based on hymns from the regular Lutheran Communion service, Bach uses as framework a 'Trinitarian' Prelude and Fugue (the '*St. Anne's*,' BWV 552): the majestic Prelude is based on three themes, the Fugue on three which are cleverly interlocked. In the Credo of the *B minor Mass* Bach uses the word *Credo* 7 x 7 = 49 times, *in unum Deum* 7 x 12 = 84 times; at the end of the fugue *Patrem omnipotentem* Bach significantly adds the number of bars (84).[45] He not only wrote his own name by using the German note names B-A-C-H (B flat-A-C-B) but, in the tradition of Werckmeister, Picander and J.J. Schmidt, did so by means of number cabalistics (as Friedrich Smend[46] has shown by extensive analysis and research). In the *Orgelbüchlein* a combination of symbols, allegory, imitation and semantics of number is used: allegory in the form of associative motifs, e.g., 'rocking-,' 'fall-' and 'damnation-,' 'ascension-' and 'descension-' motifs; imitation by means of 'knocking-,' 'trembling-,' 'trumpet-' motifs; semantics of number (e.g., twelve to symbolize the Old Year in *Das alte Jahr vergangen ist*); symbolism, e.g., strict imitation or canon as symbol of 'following,' 'obedience,' 'bondage,' 'captivity.'

Significant for an understanding of Bach's music is that, no matter how spontaneous or inspired it sounds, behind the seemingly musical construction and the psychological aspect of mood creation, there exists a structure of a more abstract nature. The two levels of communication, the affective and more abstract symbolism, usually complement one another. A rendering of his music should therefore not be restricted to a purely emotional level, or become a purely abstract reflection of a rational art. It should recognize this kind of Baroque *Gesamtkunstwerk* (total, integrated work of art), which relied on century-old concepts, and which is rooted in Bach's musical philosophy. That Bach was attacked during the late 1730's by the younger Johann Adolph Scheibe,[47] a fervent supporter of the new purely naturalistic aesthetics of the Enlightenment, is evidence of Bach's traditional ideas. In the spirit of the new generation he denounced Bach's symbolism and metaphysical emphasis and urged him to follow the new naturalistic ideal, to express his own natural emotions in an uncomplicated musical style, instead of reflecting a speculative, transcendental world or creating a prescribed mood by means of complex music.

Two short examples of Easter music from the *Orgelbüchlein* may serve to illustrate the combination of mood and more abstract symbolism in his chorale preludes:

1. Erschienen ist der herrliche Tag

'The glorious day has come, for which no man can be joyful enough: Christ, our Lord triumphs today and carries with him all his enemies in captivity. Alleluia!' The metaphor is the one used by the apostle Paul to portray Christ's triumph over death and damnation. The enemy, in accordance with Roman custom, is tied to the victor's chariot for a triumphant procession through the city. Bach unites symbolism and mood very effectively. As symbol he uses a canon (strict voice imitation) between the upper voice and the bass. In effect the bass is 'tied' to the upper voice and 'dragged' along at a fixed distance. The metaphor of the 'captive' enemy becomes completely clear in this context. Simultaneously the basic mood of triumph is created by the use of a pompous, anapestic rhythm-motif (short-short-long) in the central voice parts and the triumphant climax in the change to D major (tierce de Picardie) in the final cadence (Ex. 30).

Ex. 30

This example clearly illustrates the difference in communication level between direct and indirect symbols. While the triumphant mood of the music is directly experienced by the listener, the symbol of 'capitivity' requires familiarity with and decoding within the Lutheran theology. The same applies to the next example.

2. Jesus, Christus, unser Heiland

'Jesus Christ, our Redeemer, who has triumphed over death, is risen and has taken death into capitivity. Alleluia!' The celebrative mood is captured by the subdivision of the 4/4- into 12/8-metre by means of triplets. A change from basic binary to triple note values was often used before Bach (by Schütz, e.g.) to capture, portray and symbolize something of the blissfulness and joy of the heavenly, ethereal and perfect. Apart from the basic mood and the use of semantics of number, the resurrection itself is portrayed allegorically by a dynamically ascending, leaping interval. This resurrection-motif of two ascending notes occurs in all the voice parts and in various intervals ranging from the second to the full octave. It is most effective because of the emphasis on the second, 'off-beat' note which is tied to the next group of notes (Ex. 31).

Ex. 31

In conclusion, recognition of the rhetorical and other symbolic aspects of Bach's organ music is not only a matter of importance to the music historian. For the listener it means a keener perception of Bach's musical idiom as well as added enjoyment; for the performer, an appropriately rational-affective rendering with its implications for articulation, ornamentation, choice of timbre, dynamics, tempo and agogics.

JACOBUS KLOPPERS
The King's College, Edmonton

Addendum I

Some of the approximately 82 different musical tropes and figures developed from the sixteenth to eighteenth centuries:

Grammatical and pictorial figures:

Anabasis:	Ascending scale (e.g., 'ascension of Christ')
Circulatio:	Circulating movement (e.g., revolving movement, 'encirclement')
Katabasis:	Descending scale (e.g., 'descent,' 'depression,' 'humiliation')
Hypotyposis:	Tone painting (imitation in music of sounds, movements, gestures of specific qualities of an object)
Passus duriusculus:	A dissonant melodic progression (e.g., the tritone)
Saltus duriusculus:	Dissonant interval (e.g., diminished and augmented intervals)

Affective figures:

Anaphora (Repetition)	and other kinds of repetition: Extended (Paronomasia), emphatic (Epanalepsis), varied (Variatio), divergent (Polyptoton), with similar phrase endings (Epistrophe)
Antitheton:	Antithesis (any musical contrast, e.g., inversion of a motif)
Aposiopesis:	General pause, silence (to depict death, sighing, deep sorrow, awe)
Congeries:	Accumulation of the same idea

Dialogismus:	Musical dialogue (e.g., the antiphonal style, concerto grosso, organ dialogues)
Distributio:	'Analysis' of the main idea, the fractionation of a theme into small motifs
Dubitatio:	Expression of doubt and uncertainty (e.g., dubious modulation, sudden standstill)
Ellipsis:	Ellipse, a 'wrong conclusion' (illogical progression, disruption of an idea, deceptive cadence)
Emphasis:	Emphasis (accent, an emphatic chord)
Exclamatio:	Musical exclamation, e.g., by means of a note or chord that is not sustained but broken off
Gradatio, Climax:	Musical climax, e.g., by parallel ascending voices
Hyperbaton:	'Wrong order' of notes, 'distortion' of a theme or motif because of excitement
Hyperbol:	Hyperbole, exaggeration in music
Interrogatio:	Question or rhetorical question (e.g., an ascending appogiatura in a half-cadence)
Metalepsis:	Premature conclusion (premature entry of a voice or motif)
Mimesis:	Ridicule in music (e.g., repetition in a dialogue at a different pitch level)
Parrhesia:	Harsh expression of rage, conflict, anguish, pain by means of dissonants and dissonant chord-progressions
Pleonasmus:	A redundant repetition of an idea because of emotional involvement
Polysyndeton:	Postponement of the final cadence
Tmesis:	Disruption of a melodic line by inserting rests (as an expression of deep sorrow and sighing)

Addendum II

The educational system in Germany during the Baroque period:

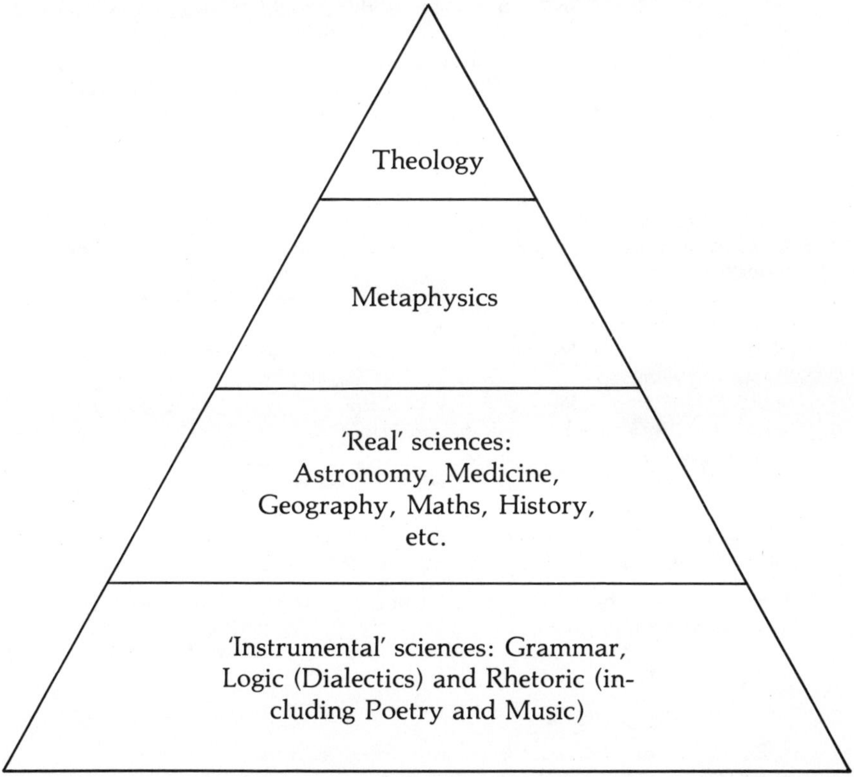

Notes

1 Research on musical rhetoric and symbolism was pioneered by German musicologist Arnold Schering at the beginning of this century. What had been earlier considered as Bach's unique musical vocabulary for conveying textual ideas (by, e.g., Albert Schweitzer), proved to be a 'vocabulary' of musical rhetoric widely used since the sixteenth century, and of an ancient musical symbolism still alive in Bach's time. More than 200 treatises from the sixteenth to the eighteenth century provided a theoretical basis for subsequent musical analyses by Schering and others. A complete list of published research on this subject is beyond the scope of this paper. For Bach, cf., e.g., Albert Schweitzer, *Johann Sebastian Bach* (Leipzig, 1908; reprint Wiesbaden: Breitkopf und Härtel, 1960); Rudolf Wustmann, 'Tonartensymbolik zu Bachs Zeit,' in *Bach-Jahrbuch* (1911), pp. 60 ff.; Arnold Schering, 'Die Lehre von den musikalischen Figuren im 17. und 18. Jahrhundert,' in *Kirchenmusikalisches Handbuch* (1908), pp. 108ff.; also, 'Bach und das Symbol,' in *Bach Jahrbuch* (1937), pp. 83-95; Hans-Heinrich Unger, *Die Beziehungen zwischen Musik und Rhetorik im 16.-18. Jahrhundert* (Würzburg: Konrad Triltsch, 1941; reprint Hildesheim: George Olms, 1969); Friedrich Blume, 'J.S. Bach,' in *Die Musik in Geschichte und Gegenwart*, vol. 1 (1949-51), pp. 962ff.; Arnold Schmitz, *Die Bildlichkeit der wortgebundenen Musik J.S. Bachs* (Mainz: B. Schott, 1950); Jacobus Kloppers, *Die Interpretation und Wiedergabe der Orgelwerke Bachs: Ein Beitrag zur Bestimmung von Stilgerechten Prinzipien* (Ph.D. Dissertation, J.W. Goethe University, Frankfurt/Main, 1965; distributed by Bärenreiter Antiquariat, Kassel); Timothy Albrecht, *Musical Rhetoric in Selected Organ Works of Johann Sebastian Bach* (Dissertation, University of Rochester, 1978); Warren Kirkendale, 'Ciceronians versus Aristotelians on the Ricercar as Exordium, from Bembo to Bach,' *Journal of the American Musicological Society*, 32 (1979), 1-44; Ursula Kirkendale, 'The Source for Bach's "Musical Offering": The "Institutio Oratoria" of Quintilian,' *Journal of the American Musicological Society*, 33 (1980), 88-141.

2 Cf. Heinrich Lausberg, *Handbuch der literarischen Rhetorik: Eine Grundlegung der Literaturwissenschaft* (Munich: M. Hueber, 1960), pp. 140ff.; Edward P.J. Corbett, *Classical Rhetoric for the Modern Student* (New York: Oxford University Press, 1971), pp. 3ff.; Unger, *op. cit.*, p. 3.

3 *Ibid.*, pp. 17-20.

4 Athanasius Kircher, *Musurgia Universalis* (Rome: Francesco Corbelletti, 1650; reprint Hildesheim: George Olms, 1970), vol. 1, liber 5, cap. 19.

5 Johann Mattheson, *Der vollkommene Capellmeister* (Hamburg: Christian Herold, 1739), pp. 121-32.

6 *Ibid.*, pp. 235-36.

7 Cf. Unger, pp. 99-109; Kloppers, pp. 126-75.

8 Cf. Johann Mattheson, *Das Neu-Eröffnete Orchestre* (Hamburg: Schillers Erben, 1713), pp. 232-51; also Rita K. Steblin, *Key Characteristics in the eighteenth and early nineteenth centuries: A historical approach* (Thesis, University of Illinois at Urbana-Champaign, 1981), pp. 63-86.

9 Johann Kirnberger, *Die Kunst des reinen Satzes* (Berlin and Königsberg: Decker & Hartung, 1774-1779), Vol. 2, pp. 102-104; Johann Joachim Quantz, *Versuch einer Anweisung die Flute traversière zu Spielen* (Berlin, 1752; reprint Kassel: Barenreiter, 1953), chapter 11 §16.

10 Kirnberger, vol. 2, pp. 105ff.; Quantz, chapter 11 §16; Mattheson, *Capellmeister*, pp. 160-70.

11 Kirnberger, p. 106; Friedrich W. Marpurg, *Anleitung zum Clavierspielen* (Berlin: Haude und Spener, 1765), p. 16.

12 Friedrich W. Marpurg, *Historisch-Kritische Beyträge zur Aufnahme der Musik* (Berlin: Schützens Witwe, 1757), p. 120.

13 Cf. Unger, pp. 63-96 for a complete list and description of these sources; also Arnold Schmitz, 'Figuren, musikalisch-rhetorische,' in *Die Musik in Geschichte und Gegenwart*, ed. Friedrich Blume, 4 (Kassel: Bärenreiter, 1955), 176-83.

14 Cf. Unger, pp. 112-18.

15 Cf. Friedrich Blume, *Renaissance and Baroque Music: A Comprehensive Survey* (translation of two of Blume's German articles by M.D. Herter Norton; New York: W.W. Norton, 1967), pp. 89-92.

16 Joseph Müller-Blattau, *Das Verhältnis von Wort und Ton in der Geschichte der Musik* (Stuttgart: Metzler, 1952), p. 12.

17 Kurt Huber, *Musikästhetik* (Ettal: Buch-Kunstverlag, 1954), pp. 45-47; Unger, p. 24.

18 Cf. also Müller-Blattau, p. 23.

19 Cf. Blume, *Renaissance and Baroque Music*, pp. 90-93.

20 Cf. *ibid.*, p. 89.

21 Cf. also Egert Pöhlmann, 'Antikenverständnis und Antikenmissverständnis in der Operntheorie der Florentiner Camerata,' *Die Musikforschung* (1969), pp. 5-13.

22 Cf. Kurt Gudewill, 'Heinrich Schütz,' *Die Musik in Geschichte und Gegenwart*, ed. Friedrich Blume, 12 (Kassel: Bärenreiter, 1965), pp. 213-14; H.H. Eggebrecht, *Heinrich Schütz: Musicus Poeticus* (Göttingen: Van den Hoeck und Ruprecht, 1959); H.H. Eggebrecht, 'Zum Figur-Begriff der Musica Poetica,' *Archiv für Musikwissenschaft*, 16 (1959), 57-69.

23 Cf. Adelheid Beckmann, *Motive und Formen der deutschen Lyrik des 17 Jahrhunderts und ihre Entsprechungen in der französischen Lyrik seit Ronsard* (Tübingen: M. Niemeyer, 1960), pp. 20, 30, 89-90, 130-31; cf. also Erich Trunz, 'Weltbild und Dichtung im deutschen Barock,' *Aus der Welt des Barock*, ed. Richard Alewyn (Stuttgart: Metzler, 1957), pp. 19-20, 30-31.

24 Cf. Mattheson, Capellmeister, chapter 13 §61, p. 103 §29; Johann Mattheson, *Critica Musica*, vol. 1 (Hamburg: im Selbstverlag, 1722), p. 199.

25 For a comprehensive list, cf. Unger, pp. 156-159 and Schmitz, 'Figuren,' pp. 176-77.

26 Verbal rhetoric underwent in itself a classicist transformation in Germany after 1700. Under the influence of Leibniz, Thomasius, Hallbauer, Gottsched and others it assumed greater clarity, directness and simplicity. Cf. Ursula Stötzer, *Deutsche Redekunst im 17. und 18. Jahrhundert* (Halle/Saale: M. Niemeyer, 1962), pp. 73-77, 90, 184-203.

27 Cf. Blume, *Renaissance and Baroque Music,* pp. 111-12.

28 Mozart wrote in 1781: 'In the opera poetry should become the obedient daughter of music. ...' Cf. Friedrich Blume, 'Wolfgang Amadeus Mozart,' *Die Musik in Geschichte und Gegenwart,* ed. Friedrich Blume, 9 (Kassel: Bärenreiter, 1961), 791, 795.

29 Cf. Friedrich Blume, 'Die Musik des Barock in Deutschland,' *Syntagma Musicologicum,* ed. Martin Ruhnke (Kassel: Bärenreiter, 1963), pp. 195-96; Friedrich Blume, 'J.S. Bach,' *Die Musik in Geschichte und Gegenwart,* ed. Friedrich Blume (Kassel: Bärenreiter, 1949), p. 103; Walther Blankenburg, 'Bach und die Aufklärung,' *Bach-Gedenkschrift,* ed. Karl Matthaei (Zürich: Atlantis, 1950), p. 31.

30 Cf. Blume, 'Die Musik des Barock in Deutschland,' p. 195.

31 Printed in Philipp Spitta, *J.S. Bach,* 5th edition (Wiesbaden: Breitkopf und Härtel, 1962), vol. 2, pp. 915-16.

32 Cf. the controversy following Blume's 'new Bach concept': Friedrich Blume, 'Umrisse eines neuen Bach-Bildes,' *Musica* (1962), pp. 169-76.

33 Johann Nicolaus Forkel, *J.S. Bach: Leben, Kunst und Kunstwerke* (Leipzig: Hoffmeister und Kühnel, 1802), pp. 24-25.

34 Birnbaum mentioned this in his defence of Bach against Scheibe, who favoured the new eighteenth century rationalistic and naturalistic tendencies in rhetoric and musical composition. Bach's musical rhetoric, which included various archaic tropes and figures and a complex religious symbolism, was unacceptable to Scheibe. Cf. Johann Adolph Scheibe, *Critischer Musikus,* 2nd edition (Leipzig: Breitkopf, 1745), p. 997.

35 Cf. Blume, 'J.S. Bach,' p. 1028.

36 Cf. Wilibald Gurlitt, 'Musik und Rhetorik: Hinweise auf ihre geschichtliche Grundlageeinheit,' *Helicon Revue internationale de la literature,* 5 (1944), 69.

37 Cf. also Kloppers, pp. 78-90, 113-21.

38 Mattheson, *Critica Musica,* vol. 1, p. 199.

39 As summarized in Blume, *Renaissance and Baroque Music,* pp. 112-16; also Blume, 'J.S. Bach,' pp. 1030-31.

40 See, i.e., Huber, pp. 45-47.

41 Karl H. Wörner, *Geschichte der Musik* (Göttingen: Van den Hoeck und Ruprecht, 1965), pp. 147-149; cf. also Blume, *Renaissance and Baroque Music,* pp. 5, 6, 10.

42 Marin Mersenne, *Harmonie Universelle* (Paris, 1636; reprint edition, Paris: Centre national de la recherche scientifique, 1963).

43 Kircher, *Musurgia Universalis.*

44 Andreas Werckmeister, *Musicalische Paradoxal-Discourse oder Ungemeine Vorstellungen ...* (Quedlinburg: T.P. Calvisius, 1707).

45 Blume, 'J.S. Bach,' p. 1030.

46 Cf. especially Friedrich Smend, *J.S. Bach bei seinem Namen gerufen: Eine Noteninschrift und ihre Deutung* (Kassel: Bärenreiter, 1950).

47 Scheibe, p. 890.